Eat or be Eaten
Predator Sensitive Foraging Among Primates

Edited by
Lynne E. Miller
MiraCosta College, Oceanside, California

CAMBRIDGE
UNIVERSITY PRESS

PUBLISHED BY THE PRESS SYNDICATE OF THE UNIVERSITY OF CAMBRIDGE
The Pitt Building, Trumpington Street, Cambridge, United Kingdom

CAMBRIDGE UNIVERSITY PRESS
The Edinburgh Building, Cambridge CB2 2RU, UK
40 West 20th Street, New York, NY 10011-4211, USA
477 Williamstown Road, Port Melbourne, VIC 3207, Australia
Ruiz de Alarcón 13, 28014 Madrid, Spain
Dock House, The Waterfront, Cape Town 8001, South Africa

http://www.cambridge.org

First published 2002

Printed in the United Kingdom at the University Press, Cambridge

Typeface Hollander 9.5/13.5 pt   *System* QuarkXPress™   [SE]

A catalogue record for this book is available from the British Library

Library of Congress Cataloguing in Publication data

Eat or be eaten: predator sensitive foraging among primates / Lynne E. Miller,
editor.
   p. cm.
Includes bibliographical references.
ISBN 0 521 80451 5 – ISBN 0 521 01104 3 (pb)
1. Primates – Behavior.   2. Primates – Food.   3. Predation (Biology). I. Miller,
Lynne E. (Lynne Elizabeth), 1963–
QL737.P9 E24 2002
599.8′153–dc21   2001037947

ISBN 0 521 80451 5 hardback
ISBN 0 521 01104 3 paperback

Contents

Contributors

SIMON K. BEARDER
Department of Anthropology
Nocturnal Primate Research Group
School of Social Sciences and Law
Oxford Brookes University
Oxford OX3 0BP
United Kingdom
skbearder@brookes.ac.uk

JÚLIO CÉSAR BICCA-MARQUES
PUCRS/Faculdade de Biociências
Av. Ipiranga, 6681 Prédio 12A
Caixa Postal 1429
Porto Alegre, RS 90619-900
Brazil

HANNAH M. BUCHANAN-SMITH
Scottish Primate Research Group
Department of Psychology
University of Stirling
Stirling, FK9 4LA
Scotland, United Kingdom

COURTNEY A. BUZZELL
Department of Anthropology
Kent State University
Kent, OH 44242
USA

NANCY G. CAINE
Department of Psychology
California State University, San Marcos
San Marcos, CA 92096
USA
ncaine@csusm.edu

MARINA CORDS
Anthropology Department
Columbia University
New York, NY 10027
USA
mc51@columbia.edu

GUY COWLISHAW
Zoological Society London
Institute of Zoology
Regents Park
London, NW1 4RY
United Kingdom
guy.cowlishaw@ioz.ac.uk

ANTHONY DI FIORE
New York Consortium in
    Evolutionary Primatology
Department of Anthropology
New York University
Rufus Smith Hall, Rm 801
25 Waverly Place
New York, NY 10003
USA
anthony.difiore@nyu.edu

KARIN L. ENSTAM
Department of Anthropology
University of California, Davis
Davis, CA 95616
USA

PAUL A GARBER
Department of Anthropology
University of Illinois
109 Davenport Hall
607 S. Mathews Ave.
Urbana, IL 61801
USA
p-garber@uiuc.edu

TERRENCE M. GLEASON
Department of Anthropology
Washington University
St. Louis, MO 63130
USA

RUSSELL A. HILL
Department of Anthropology
University of Durham
43 Old Elvet
Durham, DH1 3HN
United Kingdom
R.A.Hill@durham.ac.uk

LYNNE A. ISBELL
Department of Anthropology
University of California, Davis
Davis, CA 95616
USA
laisbell@ucdavis.edu

LYNNE E. MILLER
Department of Anthropology
MiraCosta College
One Bernard Drive
Oceanside, CA 92056
USA
lmiller@yar.miracosta.cc.ca.us

K.A.I. NEKARIS
Department of Anthropology
Washington University
St. Louis, MO 63130
USA

MARILYN A. NORCONK
Department of Anthropology and
    School of Biomedical Sciences
Kent State University
Kent, OH 44242
USA
mnorconk@kent.edu

DEBORAH J. OVERDORFF
Department of Anthropology
University of Texas – Austin
Austin, TX 78712
USA
overdorff@mail.utexas.edu

MARK J. PRESCOTT
RSPCA
Wilberforce Way
Southwater, Horsham
West Sussex, RH13 7WN
United Kingdom

MICHELLE L. SAUTHER
Department of Anthropology
University of Colorado, Boulder
Boulder, CO 80309
USA
sauther@stripe.colorado.edu

RYAN G. SELTZER
Department of Anthropology
Tulane University
New Orleans, LA 70118
USA

VOLKER SOMMER
Department of Anthropology
University College London
Gower Street
London WC1E 6BT
United Kingdom
V.Sommer@ucl.ac.uk

ELISABETH H. M. STERCK
Ethology and Socioecology
Utrecht University
Padualaan 14
P.O. Box 80086
3508 TB Utrecht
The Netherlands
e.h.m.sterck@bio.uu.nl

SUZANNE G. STRAIT
Department of Biological Sciences
Marshall University
Huntington, WV 25755
USA

ADRIAN TREVES
Department of Zoology
University of Wisconsin, Madison
250 North Mills St.
Madison, WI 53706
USA
atreves@facstaff.wisc.edu

NICOLA L. UHDE
Department of Anthropology
University College London
Gower Street
London WC1E 6BT
United Kingdom

Preface

The story of this volume is probably a common one. The book was born out of my frustration over an apparent impasse in primatology and my excitement at discovering that other ecologists had, in fact, moved beyond that obstacle. From my first year in graduate school, I was frustrated by what I perceived as a false dichotomy, the seemingly endless debate among primate socioecologists over the adaptive functions of grouping: Did primates live in groups to reduce predator pressure or to increase foraging success? Surely, I always thought, both factors could be important. Then I got out to the field and was fascinated by the variability in behavior that I was observing among capuchin monkeys. Some monkeys took risks that others avoided, and these different strategies led to differential access to food and water. Clearly, the monkeys were behaving in ways that balanced their simultaneous desires for safety and full stomachs, and the particular balance that each individual struck was based (it seemed to me) upon how vulnerable that individual was (or perceived itself to be). Upon delving into the ecological literature, I found a vast body of data pertaining to this exact balancing act: predator sensitive foraging. Countless scientists working with diverse taxa – insects, fish, birds, rodents, ungulates – were already investigating the strategies that animals use to enhance foraging success under the constraints of predator pressure. But where was the research on primates? There were some important studies, but they were remarkably few.

My enthusiasm for exploring predator sensitive foraging among primates led me to propose a symposium on the subject. I began to get in touch with friends and colleagues who were working on different aspects of primate ecology but everyone was too busy to participate in such a symposium.[1] Someone suggested that, instead, I put the material together as an edited volume, and I was just naive enough to try it. Here is the final product.

A project like this comes to fruition because of the diligent, cooperative efforts of many people (and how anyone ever completed such a task without email is impossible for me to imagine). First and foremost, I would like to thank the authors for their countless hours of writing, reviewing and revising manuscripts, not to mention the years and years of field and laboratory research that is presented here. I am also grateful for periodic scholarly exchange with Anthony Di Fiore, Stephen Miller, Marilyn Norconk, and Adrian Treves. I want simultaneously to thank and to curse Paul Garber for encouraging me to undertake this project in the first place. I also deeply appreciate the staff of Cambridge University Press (especially Tracey Sanderson, Sarah Jeffery, Miranda Fyfe and Nicola Stearn) for answering dozens of questions, for prodding me along, and for having faith in me and in this volume. Finally, and most importantly, I would like to thank my family. Editing this volume has taken a lot of my time, and therefore a lot of theirs, too. I would not have come so far without their constant love and support. I dedicate this volume to *all* of my family, but especially to Arthur Edward and Arthur Peterson, two great sources of inspiration.

Lynne E. Miller

San Diego, California, May 2001

[1] That symposium later took place, at the Eighteenth Congress of the International Primatological Society in Adelaide, Australia, in January, 2001, but the volume was well under way by that time. However, hearing these authors present their material, back-to-back, demonstrated the importance of this body of information and provided real inspiration to complete this project. I am grateful to those who participated in that symposium: S.K. Bearder, J.C. Bicca-Marques, H.M. Buchanan-Smith, M. Cords, N.G. Caine, R.A. Hill, L.A. Isbell, M.A. Norconk, and D.J. Overdorff.

1 • An introduction to predator sensitive foraging

LYNNE E. MILLER

Studies of predator sensitive foraging in nonhuman primates are in their infancy. Predator sensitive foraging (or threat sensitive foraging, Helfman 1989) represents the strategies that animals employ to balance the need to eat against the need to avoid being eaten. For decades, these trade-offs have been investigated, in both natural and laboratory settings, by dozens of scientists working with a wide array of invertebrate, piscine, avian, and mammalian species. Ecologists working with nonprimate taxa have developed theoretical models of considerable sophistication and have generated elegant data to test those models. However, only recently have primatologists begun to turn their attention to this area of research.

The fact that studies of primates lag behind those of other taxa is due, in part, to the difficulties of observing predation on primates or even evaluating the extent to which predation imposes selective pressures on primate populations. Furthermore, field research imposes additional challenges in controlling certain variables so as to isolate and assess the effects of others. However, primate research reveals complex interactions among variables and may therefore provide a more realistic portrait of animal ecology than do over-simplified laboratory experiments. Furthermore, given the diversity in primate morphology, social structure, and habitats, this taxon represents an important opportunity to test theoretical models. It is also valuable to study the principles of predator sensitive foraging in a group of animals that rely heavily upon learning and are remarkably flexible in their behavior, as these factors may lead to further diversity and specificity in foraging strategies. Altogether, primatology has the potential to make a significant contribution to this field of research.

This volume brings together primary data from a variety of primate species living in both natural habitats and experimental settings. The objectives are (a) to demonstrate that predator sensitive foraging is relevant to a wide array of primates, of various body sizes and group sizes, and living in different habitats; (b) to explore the variables that play a role in predator sensitive foraging decisions; and (c) to engender discussion about these data, including their strengths and weaknesses, in hopes that such discussion might lead to further research. Thus, in some ways this volume represents a work in progress as we hope it will result in greater interest and further investigation into this topic.

This introductory chapter serves several purposes. First, it discusses some of the variables that affect an individual's vulnerability to predation. These variables give rise to a 'perceived' level of risk that should, in turn, influence foraging decisions as animals make adaptive compromises between avoiding predators and meeting nutrient requirements. Second, this chapter also reviews some of the relevant ecological literature. Information about primates is not included here; instead, I have left it to the authors of each chapter to review the studies that pertain directly to their subject species and/or the theoretical issues at hand. Neither does this chapter provide an exhaustive review of the nonprimate literature, as there are literally hundreds of published studies and a complete listing would be far beyond the scope of this volume. Instead, I make reference to just a few of the key articles; should the reader wish to delve deeper into this literature, this chapter provides a starting point. Finally, this introduction summarizes the contents of the volume and the contribution that each chapter makes to the overall theme.

Vulnerability, perceived risk and foraging decisions

Vulnerability is used here as a qualitative measure of the probability that an individual will be the victim of a predator at any given moment. Through experience and/or genetic inheritance, an individual animal's behavior reflects an apparent 'perceived' risk. In particular, an individual's foraging decisions are expected to vary with its vulnerability (cf., Krebs 1980, Mangel and Clark 1986, McFarland 1977, Sih 1980; see Lima and Dill 1990 for a recent review). That is, individuals who perceive themselves to be more susceptible to predation would be expected to take fewer risks when foraging, even at the cost of reduced access to important resources, while those who perceive themselves to be less susceptible exploit riskier settings in order to increase nutrient intake.

These consequent differences in foraging success are expected to have fitness ramifications (cf., Wilson 1975).

An individual's level of vulnerability depends upon a number of variables. For heuristic purposes, I have chosen to sort these variables into three categories: biological, social, and environmental. Similarly, I have organized the volume around these categories. Some chapters might arguably fit into more than one section of the volume. For example, a study that compares two species may explore the biological differences between them, the habitat preferences of each one, and individual foraging decisions based upon social rank. I have placed each chapter where the research gives greatest emphasis. Primates show considerable flexibility in their responses to selective pressures, and readers are encouraged to show similar flexibility in their use of this volume's organization.

Biological variables

By biological variables, I refer to characteristics of the species that are largely, or perhaps completely, under genetic control. For example, an individual's vulnerability might be influenced by body size which should, in turn, determine the risks it takes while foraging . (For a recent review, see Lima and Dill 1990.) From this hypothesis, we would predict that members of a larger species would be safer than members of a similar but smaller species in the same environment. Furthermore, within a single species, we would expect adults to be safer than juveniles or infants, and – for sexually dimorphic species – males should be safer than females. Again, those that are less vulnerable may exhibit riskier foraging behavior if there is incentive to do so. Body size has proven to play a large role in foraging decisions for various fish (Bishop and Brown 1992, Helfman 1989, Johnsson 1993, Werner *et al.* 1983), aquatic insects (Sih 1982), gastropods (Rochette and Himmelman 1996), crustaceans (Stein and Magnuson 1976, Wahle 1992), rodents (Brown *et al.* 1988, Kotler 1984), and ungulates (Berger and Cunningham 1988, and see Kie 1999 for a review). In at least some cases (e.g., Werner *et al.* 1983), the fact that smaller animals adjusted foraging location and/or time in order to reduce predator exposure was shown to have feeding consequences that could impact fitness. Thus, the predator sensitive decisions that individuals make can represent adaptive compromises between opposing selective pressures.

Apart from body size, some species are better equipped than

others to detect, flee from, or repel predators. For example, among rodent species, the type of habitat commonly used (e.g., habitats that offer heavy vs. light cover) varies with the volume of the auditory bulla (which probably influences these animals' ability to hear predators) and mode of locomotion (which affects the ability to escape from predators) (Brown *et al.* 1988, Kotler 1984). Stream insects differ in their predator sensitivity based upon the presence or absence of protective equipment (Kohler and McPeek 1989). Thus, the genetic endowment of each species will impact the extent to which their foraging is predator sensitive.

Taking a different (though still biological) approach, those individuals whose fitness is most heavily dependent upon nutrient intake might be expected to take the greatest risks to enhance access to resources. Thus, for many taxa, females are expected to maximize foraging effort even at the cost of greater exposure to predators. Abrahams and Dill (1989) showed that, for guppies, males needed significantly more incentive than females to forage in a risky area. Thus, the sex of an individual can impose conflicting needs. Females of many species are smaller than males, and therefore are more vulnerable, but their fitness demands may encourage them to take greater risks in order to maximize access to resources.

Following this logic, an individual's vulnerability may play a lesser role in foraging decisions than its energetic needs. For example, hungrier animals are expected to take greater risks than those that are well fed. Hunger was shown to be negatively correlated with predator sensitivity in fish (Dill and Fraser 1984, Godin and Sproul 1988, Gotceitas and Godin 1991; see also Giles 1983), aquatic insects (Kohler and McPeek 1989), and gastropods (McKillup and McKillup 1994). A few researchers have attempted to quantify the effect of hunger on predator sensitivity. Kennedy *et al.* (1994) have shown that, for bullies, the nutrient reward would have to be increased 28-fold to overcome the threat of a predator (see also Abrahams and Dill 1989). Thus, predators represent considerable disincentive to feed in richer patches, but eventually an individual's energetic state may drive it to take this risk.

Taken together, we can see that an individual's foraging decisions will depend upon biological factors that may be fixed throughout its lifetime (such as auditory equipment), factors that change relatively slowly (such as body size), and factors that may change over a relatively short period of time (such as hunger level). These biological variables will influence both vulnerability and incentive to feed, two states that often generate opposing needs for the individual. How the individual balances these needs, and the resulting patterns of behavior, are the subjects of recent primate research.

Among primates, many biological variables come into play. Morphological differences among species may influence responses to predators and the ability to flee once detected, which may in turn influence foraging decisions. In this volume, Bearder *et al.* examine calling and travel patterns in Mysore slender lorises and southern lesser galagos, and how these species foraged under varying conditions of moonlight. Their results show strong differences in anti-predator behaviors. When the rapidly leaping galagos detected potential predators, they frequently responded by mobbing and giving loud alarm calls. In contrast, lorises, lacking the ability to leap to safety, used no mobbing calls. However, contrary to common belief, galagos were generally more cryptic than lorises. Galagos usually reduced their travel and remained silent during periods without any moonlight, largely because small carnivores represented a significant risk when the subjects crossed the ground between trees. Despite their slower pace, lorises were actually less vulnerable than the galagos, perhaps because they made less frequent use of the ground. Similarly, their rates of travel and contact calling were independent of moonlight. The disparity in activity patterns probably reflects each species' attempt to balance selective pressures, pressures that are, in part, mediated by biological characteristics.

Sympatric species may employ different foraging strategies based upon their vulnerability. Prescott and Buchanan-Smith explore how risk taking varies among closely related tamarin species that are sympatric in the wild. They found that, in an experimental setting, saddle-backed tamarins were more willing than red-bellied tamarins to forage low in their enclosure when food rewards were held constant. This probably represents some intrinsic difference as it is similar to their foraging patterns in the wild. However, red-bellied tamarins were more predator sensitive than saddle-backed tamarins in so far as the red-bellied were quicker than the saddle-backed to take a greater risk (feed at a low box) if the reward was large, rather than stay at a safer site (feed at a high box) if the reward was small. Therefore the red-bellied were quicker to take the greater risk and therefore got the bigger food reward. It is sometimes difficult to discern the mechanisms by which different species express predator sensitivity, but perhaps selection has shaped primate populations so that divergent strategies allow coexistence of multiple species.

Within some species, individuals demonstrate differential ability to detect food and predators. In this volume, Caine describes research showing that, among Geoffroy's marmosets, those with trichromatic vision were better able to find food items when color was an available cue, but suffered some disadvantages that dichromats did not face when targets were color camouflaged. The author speculates that this polymorphism may contribute to cooperative foraging and predator detection in callitrichids.

In most primate species, females are constantly accompanied by their dependent offspring, and thus mothers are especially vulnerable, because they are encumbered and because their infants are often targeted by predators. Sterck's study of Thomas langurs indicates that this heightened vulnerability can lead to differential foraging strategies. The results clearly demonstrate that subjects only made use of the ground while foraging, and otherwise avoided what they may have perceived as a risky area. However, females with small infants were significantly less likely to forage on the ground than were those with larger or no infants. Thus, despite their greater energetic requirements, mothers may have to make foraging compromises in order to increase the safety of their dependent offspring.

Social variables

A variety of social variables should influence an individual's vulnerability to predation and therefore foraging decisions. (For a recent review, see Lima and Dill 1990.) One of the most important is group size. Members of larger groups are expected to be safer than those in smaller groups (cf., Alexander 1974, van Schaik 1983; see also Caraco *et al.* 1980, Jarman 1974), because of group vigilance (Elgar 1989, Pulliam 1973, Triesman 1975), dilution effects (Hamilton 1971), and the increased probability of successful mobbing (e.g., Altmann 1956, Curio 1978). Therefore, it might be expected that group size would influence predator related foraging decisions: members of larger groups, because of lower vulnerability, are able to exploit resources that are located in risky areas which members of smaller groups must avoid. However, a recent review of the literature has turned up few studies that address this issue directly (though see Clutton-Brock *et al.* 1999, Molvar and Bowyer 1994).

Group size may affect feeding patterns through the medium of shared antipredator vigilance. There is considerable evidence that members of smaller groups must devote more time to vigilance, usually measured by rates of scanning, while those in larger groups share the burden and thereby reduce individual load (see Elgar 1989, Lima 1995 and Treves 2000 for reviews). Since scanning probably reduces feeding time and/or efficiency (Lendrem 1983, 1984), group size should be positively correlated with individual food intake (Dehn 1990, Lima 1995, Quenette 1990, Roberts 1996). Thus, for species in which increasing group size is an effective antipredator mechanism (i.e., as opposed to species that practice cryptic behavior such as traveling and foraging solitarily) members of larger groups should achieve a predator related foraging advantage by greater access to certain resources and/or by greater foraging efficiency.[1] Members of smaller groups are expected to exhibit risk minimizing behavior which may reduce individual feeding success. In the long run, however, each individual must balance the selective pressures of its social environment in order to maximize its fitness.

In addition to group size, group composition might also be expected to affect individual vulnerability, and therefore foraging decisions. For example, if adult males played a protective role, then members of groups with a greater number of adult or subadult males (or other sentinels) might be expected to exhibit riskier behavior, especially if it afforded them greater access to important resources (e.g., in birds: Yasukawa et al. 1992; in viverrids: Clutton-Brock et al. 1999, Rasa 1986; see also Bednekoff 1997 and Lima and Dill 1990).

Cohesion and centrality are additional social variables often thought to correlate with vulnerability. Due to increased vigilance and a dilution effect, individuals with near neighbors are thought to be safer than those at a distance from conspecifics (or other potential prey items). Similarly, those in the center of the group are expected to be targeted by predators less frequently than those on

---

[1] From a different perspective, group foraging may instead have a negative impact on feeding success. It is often suggested that optimal group size represents a trade-off between the threat of predation and the cost of intragroup feeding competition (cf., van Schaik 1983, Dunbar 1988). Therefore, for some populations, members of large groups, not small groups, may be making the adaptive compromises as they endure higher rates of feeding competition, and therefore lower food intake, in exchange for safety. However, this model swings away from our focus on predator sensitive foraging and therefore will not be discussed further here.

the periphery (e.g., in ungulates: Berger and Cunningham 1988). Of course, individual spacing and location within the group are generally more labile characteristics than are group size or composition; they can be adjusted as demanded by changing levels of risk or hunger. Thus, modifying its position may be one mechanism by which an individual can reduce its vulnerability (e.g., in ungulates: Green 1992). However, feeding in proximity to conspecifics may reduce foraging success (e.g., in birds: Cresswell 1998; in ungulates: Lipetz and Bekoff 1982), and animals must therefore balance risk against nutritional needs, which is the essence of predator sensitive foraging.

Finally, social rank has been shown to affect predator sensitive foraging (e.g., in fish: Gotceitas and Godin 1991) but the interactions among variables can be complex. For example, higher ranking individuals may be larger, thus less vulnerable, and therefore willing to take risks (e.g., in fish: Johnsson 1993). Alternatively, those of lower rank may have poor access to food, and thus be driven by hunger to take more risks (e.g., in fish: Metcalfe 1986; in birds: Ekman and Askenmo 1984). Furthermore, higher ranking individuals can probably maintain choice locations within the group and therefore have greater options when it comes to developing a foraging strategy.

Although many primatologists have explored the effects of group size on feeding behaviors, few have specifically addressed predator sensitive foraging (see individual chapters for references). In this volume, Miller tests the hypothesis that membership in a larger group bestows a foraging advantage through the medium of reduced vulnerability to predators. This study of wedge-capped capuchins demonstrates that, during certain months of the year, females living in a larger group foraged and traveled on the ground three times as frequently as did those in a smaller group, and collected twice as much food from this substrate. Females in the smaller group, who restricted their foraging to arboreal locations, experienced significantly lower levels of food intake for several months of the year. Thus, among capuchins, predator sensitive foraging may have significant fitness ramifications.

Lending further support to the model that group size affects predator sensitive foraging, the chapter by Sauther explores the behavior of two troops of ring-tailed lemurs. Sauther's data show that members of a small group avoided foraging on the ground when predator pressure was high, and that this pattern resulted in reduced leaf and fruit intake. The small group was also more likely to form mixed-species associations with sifakas, especially while feeding and most frequently when infants were present. In contrast, the large group was more willing to enter new regions of the forest, which was associated with increases in both fruit feeding and predator encounters. Thus, Sauther's work confirms that predator sensitivity entails trade-offs between safety and foraging success, and that primates alter their behavior in ways that balance these variables, within the context of their social environments.

In some cases, social variables are more important than biological factors in predator related behaviors. Overdorff *et al.* explore the extent to which body size and group size influence the risks that prosimians take when foraging. Their study compares the behaviors of three species: red-bellied lemurs that are small-bodied and live in small groups, rufus lemurs that are small-bodied but live in larger groups, and Milne Edward's sifakas that are larger-bodied and also live in larger groups. The data demonstrate that the variance in these species' use of exposed habitat and distance between nearest neighbors was more heavily dependent upon group size than body size.

It has often been suggested that primates form mixed-species aggregations in order to reduce predator vulnerability by increasing the overall size of the social unit. Increased troop size, however, is generally expected to result in higher foraging costs. In this volume, Garber and Bicca-Marques explore predator sensitive foraging in single- vs. mixed-species troops of tamarins. A review of the tamarin literature reveals no consistent differences in predation risk for members of single- vs. mixed-species troops. Moreover, data from their study of mixed-species troops show no signs of cooperative vigilance, or other indications that, as the number of animals present at a feeding site increased, individuals foraged in a less predator sensitive way. Tamarins have developed other antipredator strategies, such as avoiding use of the same travel routes on consecutive days, and having certain group members (such as adult males) serve as sentinels while others are feeding. However, based on time spent at feeding sites, there was no evidence that the animal that served as a sentinel had lower foraging success. Thus, new data constantly encourage us to reconsider our models.

Predation models suggest that primate groups should increase their cohesion (or reduce spread) in response to risk. However, in this volume, Isbell and Enstam present data that support a different model. They have compared the behavior of (a) vervets and patas monkeys living in the same ecosystem, under similar threat of predation but using two different habitats, and (b) the same group of vervets while using two different subhabitats. The results suggest that group spread was more highly correlated with resource dispersion than with predator pressure. Again, some primates demonstrate limited predator sensitivity, and force us to look for alternative explanations for their behavior.

Individual primates occupying risky positions within a social group may be forced to devote time and energy to antipredator behaviors; however, this does not always have an impact on feeding efficiency. Gleason and Norconk studied predator responses among white-faced sakis and found that alarm calling was more frequent among subjects that were at the periphery of the group, and less frequent among those in close proximity to conspecifics. However, the data revealed no association with foraging success. This chapter posits that more extreme antipredator behavior, such as prolonged mobbing or cryptic 'freezing,' must eventually reduce feeding time, but these data are currently unavailable. Studies such as these generate testable hypotheses and therefore contribute to our advancement of research.

Environmental variables

An individual's risk will vary with the area in which it is foraging. It has generally been asserted that open areas leave animals more vulnerable than do areas with heavy cover (e.g., in rodents: Brown *et al.* 1988, Kotler 1984, though see Longland 1994; in birds: Suhonen 1993; in ungulates: Underwood 1982). Animals may opt for foraging substrates of lesser quality in order to increase cover and thereby reduce risk (e.g., in fish: Abrahams and Dill 1989, Gilliam and Fraser 1987, Holbrook and Schmitt 1988, Kennedy *et al.* 1994, Werner *et al.* 1983, though see Morin 1986), indicating that individuals must make adaptive compromises between the needs to maximize foraging success and minimize the threat of predation. However, other studies demonstrate that animals may increase vigilance in habitats with restricted visibility (e.g., in ungulates: Goldsmith 1990, Underwood 1982). These individuals are actually less vulnerable in open areas owing to their heightened ability to spot predators and thereby avoid attack. Thus, we may need to have detailed information on predator–prey interactions before we can make appropriate predictions about behavior under different habitat conditions.

Vulnerability varies not only with foraging location but also with the type of predators and their hunting strategies. In the presence of terrestrial predators, such as large-bodied felids, the ground is probably more risky than the trees; with aerial predators, the tree tops or terminal ends of branches are probably more risky than are regions at midcanopy or close to the trunk (e.g., in birds: Suhonen 1993). For nocturnal foragers, the degree of ambient light might influence risk, because some hunters, such as owls, are often more effective on moonlit nights than on dark nights (e.g., in rodents: Clarke 1983, Lockard and Owings 1974, Longland 1994, Wolfe and Summerlin 1989). For other hunters, such as snakes, moonlight may have little impact and prey species respond accordingly (e.g., in rodents: Bouskila 1995). Thus, an individual's vulnerability depends upon the set of predators that represent selective pressures.

An individual cannot make major changes in its habitat (e.g., the distribution of resources and predators) but it can make minor adjustments in where and when it forages so as to balance safety against access to food. For example, individuals may proceed to feed in risky habitats when other factors reduce their vulnerability (e.g., larger group size or greater ability to detect predators) (in rodents: Brown *et al.* 1988, Kotler 1984; in ungulates: Hirth 1977, Jarman 1974, Molvar and Bowyer 1994), or when their hunger levels are higher (e.g., in fish: Gotceitas and Godin 1991), or when the reward is greater (e.g., in fish: Holbrook and Schmitt 1988; in ungulates: Underwood 1982; see also Ferguson *et al.* 1988). Conversely, risk may be reduced by moving to an area that provides greater protective cover (e.g., in birds: Suhonen 1993) and nocturnal animals may restrict foraging on moonlit nights (e.g., in rodents: Brown *et al.* 1988, Longland 1994). Vulnerability may be influenced by various biological, social and environmental variables, but many of these are beyond the individual's capacity to change. The choice of foraging substrate and timing represents one manifestation of a predator sensitive foraging strategy.

When a single primate species inhabits two different habitat types, there is an ideal opportunity to test the influence of the environment on predator sensitive foraging. In this volume, Hill and Cowlishaw explore vigilance behavior and its social and ecological correlates for two populations of chacma baboons. Their research shows that rates of vigilance in females differed dramatically across the two populations, and that this phenomenon can be largely explained by differences in feeding activity, use of areas that provide refuge, and group size and cohesion. Once these factors have been accounted for, no population differences exist, suggesting that baboons in different populations may follow a consistent antipredation strategy that is adjusted to local ecological conditions, such as the distribution of food patches. Thus, this study makes a clear case that the environment imposes selective pressures upon individuals who must then modify their behaviors to balance the costs and benefits of feeding and antipredator activity.

Many primate species exploit diverse resources, and therefore foraging decisions may vary with the type of food that is being sought. Cords demonstrates that blue monkeys foraged more often in areas of low foliage density, and foraged in the presence of fewer conspecifics (i.e., fewer in the same tree) when in search of invertebrates. Thus, these subjects made adjustments in both social and environmental variables, in directions that are expected to increase risk, in order to increase access to a certain resource. In balancing exposure to predation against nutrient demands, blue monkeys provide an excellent example of predator sensitive foraging.

Studies of vigilance usually predict that scanning will increase with exposure; that is, it is assumed that individuals are more vulnerable when they are foraging under limited cover. However, Treves examines an alternative model that may have greater explanatory value for primates. In this volume, he analyzes the behavior of redtail monkeys, red colobus monkeys, and black howler monkeys. Comparing each species across two different conditions of potential vulnerability (e.g., degree of cover, location in a tree), he found no strong relationship between risk and scanning, but did find that visual obstructions (i.e., dense foliage) were associated with increased vigilance. He concludes that the prey's ability to spot the predator may be a more important variable when it comes to estimating risk, and therefore exposed areas may actually be safer than those that offer cover.

Further data to test this model are provided by Di Fiore who evaluates how foraging behavior in woolly monkeys responds to various biological, social and ecological conditions. This study demonstrates that, in general, large-bodied woolly monkeys were not predator sensitive in their feeding activities and devoted very little time to vigilance, under any circumstances. However, rates of scanning did vary with protective cover. The data are consistent with the model that primates may decrease predator pressure by using microhabitats that allow for improved predator detection, and increase vigilance when detection is more difficult.

Species that form small groups because of feeding competition may be especially vulnerable to predators, and therefore could be expected to show selective use of microhabitats in order to reduce risk. In this volume, Uhde and Sommer argue that, for white-handed gibbons, predator pressure is not negligible but that the species probably maintains smaller (single-female) groups in order to reduce foraging costs. Under these circumstances, the authors predict a relationship between activity and height in the canopy. Their data show that gibbons used lower substrates while foraging, presumably to increase access to important resources, but chose higher substrates when resting or grooming. Although in some ways these subjects showed limited predator sensitivity, exposure to a tiger led one group to increase their cohesion and reduce feeding activity, behaviors that would be predicted by these ecological models.

•

Virtually every chapter in this volume points out the paucity of data on predator–prey interactions for primates in the wild, and the difficulties of collecting such data. Given these challenges, it should come as no surprise that the research presented in these pages exhibits great diversity and creativity in approach, as the authors have endeavored to make observations that would shed light on predator sensitive foraging among primates. Key elements are sometimes missing. Rarely is a study able to provide hard data pertaining to the relative threat of predation experienced by different subjects, and the behavioral strategies that those individuals employ to reduce risk, and the consequential impact on feeding success. Instead, most studies are forced to infer these factors from other aspects of behavior. However, taken as a whole, this body of data represents significant progress in our understanding of how primate populations have evolved under the selective pressures of predation and resource demands.

REFERENCES

Abrahams, M.V., and Dill, L.M. (1989). A determination of the energetic equivalence of the risk of predation. *Ecology* **70**: 999–1007.

Alexander, R.D. (1974). The evolution of social behavior. *Annual Review of Ecology and Systematics* **5**: 324–83.

Altmann, S.A. (1956). Avian mobbing behavior and predator recognition. *Condor* **58**: 241–53.

Bednekoff, P.A. (1997). Mutualism among safe, selfish sentinels: A dynamic game. *American Naturalist* **150**: 373–85.

Berger, J., and Cunningham, C. (1988). Size-related effects on search times in North American grassland female ungulates. *Ecology* **69**: 177–83.

Bishop, T.D., and Brown, J.A. (1992). Threat-sensitive foraging by larval threespine sticklebacks (*Gasterosteus aculeatus*). *Behavioral Ecology and Sociobiology* **31**: 133–8.

Bouskila, A. (1995). Interactions between predation risk and competition: A field study of kangaroo rats and snakes. *Ecology* **76**: 165–78.

Brown, J.S., Kotler, B.P., Smith, R.J., and Wirtz, W.O. II (1988). The effects of owl predation on the foraging behavior of heteromyid rodents. *Oecologia* **76**: 408–15.

Caraco T., Martindale, S., and Pulliam, H.R. (1980). Avian flocking in the presence of a predator. *Nature* **285**: 400–1.

Clarke, J.A. (1983). Moonlight's influence on predator/prey interactions between short-eared owls (*Asio flammeus*) and deermice (*Peromyscus maniculatus*). *Behavioral Ecology and Sociobiology* **13**: 205–9.

Clutton-Brock, T.H., O'Riain, M.J., Brotherton, P.N.M., and Gaynor, D. (1999). Selfish sentinels in cooperative mammals. *Animal Behaviour* **284**: 1640–4.

Cresswell, W. (1998). Relative competitive ability changes with competitor density: Evidence from feeding blackbirds. *Animal Behaviour* **56**: 1367–73.

Curio, E. (1978). The adaptive significance of avian mobbing. *Zeitschrift fur Tierpsychologie* **48**: 175–83.

Dehn, M.M. (1990). Vigilance for predators: Detection and dilution effects. *Behavioral Ecology and Sociobiology* **26**: 337–42.

Dill, L.M., and Fraser, A.H.G. (1984). Risk of predation and the feeding behavior of juvenile coho salmon (*Oncorhynchus kisutch*). *Behavioral Ecology and Sociobiology* **16**: 65–71.

Dunbar, R.I.M. (1988). *Primate Social Systems*. Cornell University Press: Ithaca, New York.

Ekman, J., and Askenmo, C. (1984). Social rank and habitat use in willow tit groups. *Animal Behaviour* **32**: 508–14.

Elgar, M. (1989). Predator vigilance and group size in mammals and birds: A critical review of the empirical evidence. *Biological Review* **64**: 13–33.

Ferguson, S. H., Bergerud, A.T., and Ferguson, R. (1988). Predation risk and habitat selection in the persistence of a remnant caribou population. *Oecologia* **76**: 236–45.

Giles, N. (1983). Behavioral effects of the parasite *Schistocephalus solidus* (Cestoda) on an intermediate host, the three-spined stickleback, *Gasterosteus aculeatus* L. *Animal Behaviour* **31**: 1192–4.

Gilliam, J.F., and Fraser, D.F. (1987). Habitat selection under predation hazard: Test of a model with foraging minnows. *Ecology* **68**: 1856–62.

Godin, J.-G., and Sproul, C.D. (1988). Risk taking in parasitized sticklebacks under threat of predation: Effects of energetic need and food availability. *Canadian Journal of Zoology* **66**: 2360–7.

Goldsmith, A.E. (1990). Vigilance behavior of pronghorns in different habitats. *Journal of Mammalogy* **71**: 460–2.

Gotceitas, V., and Godin, J.-G.J. (1991). Foraging under the risk of predation in juvenile Atlantic salmon (*Salmo salar* L.): Effects of social status and hunger. *Behavioral Ecology and Sociobiology* **29**: 255–61.

Green, W.C.H. (1992). Social influences on contact maintenance interactions of bison mothers and calves: Groups size and nearest-neighbor distance. *Animal Behaviour* **43**: 775–85.

Hamilton, W.D. (1971). Geometry for a selfish herd. *Journal of Theoretical Biology* **31**: 295–311.

Helfman, G.S. (1989). Threat-sensitive predator avoidance in damselfish-trumpetfish interactions. *Behavioral Ecology and Sociobiology* **24**: 47–58.

Hirth, D.H. (1977). Social behavior of white-tailed deer in relation to habitat. *Wildlife Monographs* **53**: 1–55.

Holbrook, S.J., and Schmitt, R.J. (1988). The combined effects of predation risk and food reward on patch selection. *Ecology* **69**: 125–34.

Jarman, P.J. (1974). The social organisation of antelope in relation to behaviour. *Behaviour* **48**: 215–67.

Johnsson, J.I. (1993). Big and brave: Size selection affects foraging under risk of predation in juvenile rainbow trout, *Oncorhynchus mykiss*. *Animal Behaviour* **45**: 1219–25.

Kennedy, M., Shave, C.R., and Spencer, H.G. (1994). Quantifying the effect of predation risk on foraging bullies: No need to assume an IFD. *Ecology* **75**: 2220–6.

Kie, J.G. (1999). Optimal foraging and risk of predation: Effects on behavior and social structure in ungulates. *Journal of Mammalogy* **80**: 1114–29.

Kohler, S.L., and McPeek, M.A. (1989). Predation risk and the foraging behavior of competing stream insects. *Ecology* **70**: 1811–25.

Kotler, B.P. (1984). Risk of predation and the structure of desert rodent communities. *Ecology* **65**: 689–701.

Krebs, J.R. (1980). Optimal foraging, predation risk, and territory defense. *Ardea* **68**: 83–90.

Lendrem, D.W. (1983). Predation and risk in the blue tit (*Parus caeruleus*). *Behavioral Ecology and Sociobiology* **14**: 9–13

Lendrem, D.W. (1984). Flocking, feeding and predation risk: Absolute and instantaneous feeding rates. *Animal Behaviour* **32**: 298–9.

Lima, S.L. (1995). Back to the basics of anti-predatory vigilance: The group size effect. *Animal Behaviour* **49**: 11–20.

Lima, S.L., and Dill, L.M. (1990). Behavioral decisions made under the risk of predation: A review and prospectus. *Canadian Journal of Zoology* **68**: 619–40.

Lipetz, V.E., and Bekoff, M. (1982). Group size and vigilance in pronghorns. *Zeitschrift für Tierpsychologie* **58**: 203–16.

Lockard, R.B., and Owings, D.H. (1974). Seasonal variation in moonlight avoidance by bannertail kangaroo rats. *Journal of Mammalogy* **55**: 189–93.

Longland, W.S. (1994). Effects of artificial bush canopies and illumination on seed patch selection by heteromyid rodents. *American Midland Naturalist* **132**: 82–90.

Mangel, M., and Clark, C.W. (1986). Towards a unified foraging theory. *Ecology* **67**: 1127–38.

McFarland, D.J. (1977). Decision making in animals. *Nature* **269**: 15–21.

McKillup, S.C., and McKillup, R.V. (1994). The decision to feed by a scavenger in relation to the risks of predation and starvation. *Oecologia* **97**: 41–8.

Metcalfe, N.B. (1986). Intraspecific variation in competitive ability and food intake in salmonids: Consequences for energy budgets and growth rates. *Journal of Fish Biology* **28**: 525–31.

Molvar, E.M., and Bowyer, R.T. (1994). Costs and benefits of group living in a recently social ungulate: The Alaskan moose. *Journal of Mammalogy* **75**: 621–30.

Morin, P.J. (1986). Interactions between intraspecific competition and predation in an amphibian predator–prey system. *Ecology* **67**: 713–20.

Pulliam, H.R. (1973). On the advantages of flocking. *Journal of Theoretical Biology* **38**: 419–22.

Quenette, P.Y. (1990). Functions of vigilance behaviour in mammals: A review. *Acta Oecologia* **11**: 801–18.

Rasa, O.A.E. (1986). Coordinated vigilance in dwarf mongoose family groups: The watchman's song hypothesis and the costs of guarding. *Ethology* **71**: 340–4.

Roberts, G. (1996). Why individual vigilance declines as group size increases. *Animal Behaviour* **51**: 1077–86.

Rochette, R., and Himmelman, J.H. (1996). Does vulnerability influence trade-offs made by whelks between predation risk and feeding opportunities? *Animal Behaviour* **52**: 783–94.

Sih, A. (1980). Optimal foraging behavior: Can foragers balance two conflicting demands? *Science* **210**: 1041–1043.

Sih, A. (1982). Foraging strategies and the avoidance of predation by an aquatic insect, *Notonecta hoffmanni*. *Ecology* **63**: 786–96.

Stein, R.A., and Magnuson, J.J. (1976). Behavioral response of crayfish to a fish predator. *Ecology* **57**: 751–61.

Suhonen, J. (1993). Predation risk influences the use of foraging sites by tits. *Ecology* **74**: 1197–203.

Treves, A. (2000) Theory and method in studies of vigilance and aggregation. *Animal Behaviour* **60**: 711–22.

Triesman, M. (1975). Predation and the evolution of gregariousness. I. Models for concealment and evasion. *Animal Behaviour* **23**: 799–900.

Underwood, R. (1982). Vigilance behaviour in grazing African antelopes. *Behaviour* **79**: 81–107.

van Schaik, C.P. (1983). Why are diurnal primates living in groups? *Behaviour* **87**: 120–44.

Wahle, R.A. (1992). Body-size dependent anti-predator mechanisms of the American lobster. *Oikos* **65**: 52–60.

Werner, E.E., Gilliam, J.F., Hall, D.J., and Mittelbach, G.G. (1983). An experimental test of the effects of predation risk on habitat use in fish. *Ecology* **64**: 1540–8.

Wilson, E.O. (1975). *Sociobiology: The new synthesis*. Cambridge, MA: Harvard University Press.

Wolfe, J.L., and Summerlin, C.T. (1989). The influence of lunar light on nocturnal activity of the oldfield mouse. *Animal Behaviour* **37**: 410–14.

Yasukawa, K., Whittenberger, L.K., and Nielsen, T.A. (1992). Anti-predator vigilance in the red-winged blackbird, *Agelaius phoeniceus*: Do males act as sentinels? *Animal Behaviour* **43**: 961–9.

Part I • Biological variables

2 • Dangers in the night: Are some nocturnal primates afraid of the dark?

SIMON K. BEARDER, K.A.I. NEKARIS & COURTNEY
A. BUZZELL

Introduction

The vital life functions of animals dictate that they must obtain
energy, avoid becoming an energy source for other organisms and
convert energy into the next generation. Avoiding becoming an
energy source involves defense against three major categories of
threat: (a) parasites and diseases, (b) temperature stress and other
environmental hazards and (c) predators (Dunbar 1977). The interactions between these variables may be very complex. Here we
focus on the question of how the foraging patterns of nocturnal
primates may be influenced by predation risk. Data come from
long-term field studies of two species, the Mysore slender loris
(*Loris tardigradus lydekkerianus*) from India and the southern lesser
galago (*Galago moholi*) from South Africa (Fig. 2.1). Direct evidence
of predation is very rare, especially at night, so we examine three
indirect measures of risk. First, what are the likely predators in
each study area? Second, how does each species react to potential
predators at night (predator defense)? Third, to what extent do
they appear to avoid detection (crypsis)? We then address whether
or not there is evidence for an effect on food intake or foraging
decisions.

Mysore slender lorises and southern lesser galagos diverge considerably in their methods of locomotion but both species live in
relatively open thorn tree habitats that force them to cross open
spaces on the ground. Slender lorises are unable to jump and they
generally remain below 3 m and cross between trees by stretching
and bridging. They rarely descend to the ground (less than three
times per hour) where they walk quadrupedally. Lesser galagos, in
comparison, are specialized leapers, covering up to 5 m in one
bound, as well as walking and running along branches. Their activity is concentrated at heights below 4 m and they frequently cover

Fig. 2.1. Photographs depicting a juvenile Mysore slender loris (a) and two juvenile Southern African lesser galagos (b) in typical resting postures when 'parked' by the mother.

up to 60 m along the ground by means of kangaroo-like hops (up to 30 times per hour). The two study species are broadly comparable in their mating systems (multi-male/multi-female). They both spend the day either alone or in small groups of up to seven individuals and are found alone at night on approximately 80% of spot sightings (see Nekaris 2000a and Bearder 1987 for further details). Their comparison demonstrates that the ways nocturnal primates react to encounters with predators and avoid detection vary considerably, both within and between species, and not in a way that is immediately obvious from outward appearances.

This chapter concentrates on possible relationships between predation pressure and foraging patterns at night, but it is instructive to consider that the tendency to forage at night is, in itself, at least partly influenced by the threat of predation during the daytime. Two predictions follow from this: first, that nocturnal animals should seek safe retreats during daylight hours and, second, that if

Fig. 2.1. (*cont.*).

local circumstances change to reduce the dangers during the day, or increase the costs of being active at night, then individuals should shift their activity patterns accordingly.

It is, indeed, the case that those primates that are adapted to forage efficiently at night usually remain protected from predators during the day, for example, in a nest or tree hollow, beneath dense thorns or by means of cryptic coloration and camouflage. Nevertheless, they often waken when disturbed and can escape rapidly between trees if necessary. Some galago species also engage in, apparently normal, activities during the daytime – feeding, grooming, jumping and crossing open ground with as much ease as they do at night. Similar activity patterns have been observed in captivity. This capacity to be active by day or night is very well developed in a number of lemur species in Madagascar, which are described as cathemeral (Tattersall 1987). Three types of cathemerality have been noted among lemurs. In some cases, they switch from being active by day to being active at night. Alternatively, they may be day-living for part of the year and then shift to a 24-hour activity pattern, or they may maintain a 24-hour activity cycle throughout the year (Curtis and Rasmussen 2001, Overdorff and Rasmussen 1995, Rasmussen 1999). In cases where there is a change of activity pattern, there is circumstantial evidence that it coincides with a shift in the relative balance of risks and benefits associated with avoiding dangers and obtaining energy (for example, in mid-winter when food may be frozen at night).

Any attempt to assess the effects of predation pressures is fraught with difficulties. Each study species requires a different approach depending on factors such as visibility, approachability and whether or not it is possible to use radio tracking. Given the limitations of our studies in the field, our conclusions remain tentative and in need of further research and analysis.

Methods

Lorises

The study of Mysore slender lorises was carried out at Ayyalur Interface Forestry Division (AIDF) in Dindigul District, Tamil Nadu, South India (77° 55′ E; 10° 04′ S). The 1km^2 study site was located in the foothills of the Eastern Ghats. One to two researchers and two field assistants directly observed lorises at close range for 1173 hours from October 1997 to July 1998. Twenty focal animals as well as at least 15 unknown individuals were observed during this study. Animals were identified either by distinctive markings, particularly their facial masks and shape of the nose, and distinctive behavior patterns. The study population consisted of six adult females, five adult males, one subadult female, one subadult male, one juvenile male, two female infants, one male infant, and three infants of unknown sex. Data on general activity budgets were collected using focal-animal, instantaneous point sampling (Altmann 1974) with a 5-minute sampling interval ($n = 13717$). The categories were 'inactive,' 'travel,' 'forage,' 'feed,' 'groom,' and 'other' (Nekaris 2000a,b). Data related to diet were recorded ad $libitum$ ($n = 1240$) (Nekaris 1999, 2000a). Animals were observed from a distance of 2–5 m with the aid of hand held flashlights and headlamps. To reduce disturbance to the animals, the lights were covered with red cellophane filters (Charles-Dominique and Bearder 1979). A total of about 1400 hours was spent writing down each time a loris made a loud call, as well as collecting home range data by means of noting down where each animal was in the study site. For the purposes of this chapter, light phases of the night are considered to be whenever the moon was above the horizon and dark phases were whenever the moon had set. Each time an animal was seen in the presence of a potential predator, its behavior was noted. Results were analyzed using JMP for Macintosh and SPSS (7.0) for IBM. Significance levels were set at $p < 0.05$.

In addition, 1 month was spent recording reactions of slender loris infants to the calls of the most common potential predator, the 165-g southern spotted owlet (*Athene brahma brahma*). During 19 all-night follows of two different infants, the following variables were recorded every time an owl made a vocalization: distance of the owlet to the infant; whether the infant was exposed or unexposed in relation to foliage; and the behavior of the infant. Relevant behaviors included whether the infant altered its activity in any way and, if it was exposed, whether or not it took cover in the foliage.

Galagos

Southern African lesser galagos were studied in a more or less homogeneous belt of acacia woodland (thornveld) on the cattle ranch 'Mosdene' in the Northern Province of South Africa (28° 47′ E; 24° 35′ S). The vegetation was selected specifically to allow ample opportunities for immigration and emigration in all directions. A series of traps was placed within a 1 km² study area and the initial trapping program continued until no unmarked galagos were seen or caught. Each animal was given individually distinctive ear notches and portions of the tail hair were clipped or dyed for easy recognition by day or night. Up to 20 individuals at any one time were fitted around the waist with radio transmitter belts on separate frequencies. A total of 73 belts, having an average transmission time of 48 days, were fitted to 37 different individuals to yield over 1700 positional records during discontinuous tracking. Belts were generally removed after 2 months and the animals given a period of rest to minimize the potential adverse effects of excessive wear (minimum transmitter weight, 8 g; galago weight, 170–250 g). Regular re-trapping also facilitated population monitoring and the marking of recruits. In addition, a control-trapping program was conducted in a separate locality 1 km to the west over the same time period (Bearder and Martin 1980).

The majority of nocturnal observations consisted of continuous periods spent following selected animals with the help of radio location and red flashlight, which produced a strong tapetal reflection without disturbance. Fifty-seven all-night follows were made at intervals of 7–10 days between October 1975 and May 1977, giving a total of 576 hours of continuous observation from a distance of 3–10 m. Six nights spent following individuals that were migrating have been excluded from the analysis. Behaviors were categorized as for the slender loris study and data collected using focal-animal instantaneous point sampling, but using 10-minute, rather than 5-minute, intervals (Bearder and Martin 1980). The tree in use at the end of each interval was marked with a plastic tag and its position mapped out on the following day. This allowed accurate measurement of path length (the combined distance between all marked trees) and night range length (the furthest distance between two points of travel on one night). The presence or absence of moonlight is represented in the same way as for the study of lorises. Periods of twilight were calculated as 1.5 hours after sunset and before sunrise. A further 308 hours were spent observing galagos during random checks on the positions of all radio-tagged animals. In a separate study, continuous records of selected behaviors were recorded for individuals followed at close range for 430 hours (Bearder 1969).

Indirect evidence of the diet of one predator, the genet (*Genetta tigrina*), was provided by identification of their hair in samples collected on sticky tape placed in the vicinity of kills, as well as their habit of defecating in salt-lick trays provided for the cattle on the ranch. The trays consisted of a number of separate compartments for each type of salt, placed 1 m above the ground and protected on three sides and above to prevent spoiling by wind and rain. This proved to be an ideal situation for a high-level genet midden, since the scats deposited in each compartment were preserved for many months. Twenty liters of genet droppings were collected and subjected to detailed content analysis (Andrews and Evans 1979). Finally, tree species, height and density were sampled in 32 circular plots of 20-m diameter spaced at random over the study area for comparison with plots covering preferred feeding and sleeping zones.

Results

Potential nocturnal predators in each study area

Lorises

Other than the lorises themselves, animal densities at the site in South India were low. The only potential predators of lorises by night were small wild cats (probably *Felis chaus* or *Felis rubiginosa*), domestic cats, several different snakes (including kraits and Russel's sand boa), the barn owl (*Bubo bubo*) and the spotted owlet (*Athene brahma*). The spotted owlet is one of the most common owl species in India. They are abundant around human habitation, especially around ruins, mango topes, and village grove trees (Ali 1964, 1969). Most of what is inferred about their diet comes from a study of the little owl, *Athene noctua*, in Europe (Mikkola 1983), which indicates that between 7% and 28% of the diet consists of small mammals weighing up to 65 g. The spotted owlet would be unlikely to take an adult loris (250–350 g) but infant lorises would be easy prey targets. At birth, infant slender lorises weigh only 9 g. By the fourth week of life the mother parks infants for the entire night, at which time their body weight is about 25 g (Nekaris 2000a). Circumstantial evidence of predation was provided when a cat was seen in the study area on two occasions in 1 week. During the same week, local people chopped down several trees used as a pathway by one of the focal infants. The removal of these trees made it impossible for him to avoid movement on the ground. Within 3 days of the appearance of the cat and the disappearance of the trees, he was never seen in the study area again.

Galagos

The principal danger to galagos came from genets (*Genetta tigrina*), jackals (*Canis mesomelas*), domestic cats and several species of owl, including barn owls (*Tyto alba*), spotted eagle owls (*Bubo africanus*) and grass owls (*Tyto capensis*). Snakes were very common, including the boomslang (*Dispholidus typus*) the yellow cobra (*Naja nivea*), the black mamba (*Dendroapsis polylepis*) and the python (*Python sebae*). There is firm evidence of predation on galagos based on the discovery of carcasses. The remains of seven galagos were discovered after having been consumed by a small carnivore during the night. In three cases the radio belt was located on the ground. The hind quarters of three others were cached in the branches of a tree, with the radio transmitter still around the waist. One animal,

which had not been fitted with a radio, was also found in this condition. In each case the culprit was a genet and they all occurred
on dark and unusually windy nights when the senses of the
galagos would have been impaired. A 1-year-old male was killed on
a stormy night having just migrated during five moonlit nights
from his natal range to an unknown area some 2 km away. Content
analysis of genet feces revealed jaw fragments from two galagos in
a total of 231 bone specimens of 14 prey species (Andrews and
Evans 1979).

Reactions to potential predators at night

Lorises

Wild cats were observed in the study area only seven times, and only
on four occasions were they seen within 10 m of a loris. On two occasions, the cat passed beneath a tree where an adult loris was traveling. In both cases, the loris first stared at the cat and then moved
swiftly in the opposite direction, but with no alarm call. On one of
these occasions an infant was within 20 m of the adult loris and
when the cat passed beneath the tree in which it sat, it took no
notice. On two other occasions the cat was seen directly below a tree
that contained a loris. In the first instance, they were staring
intently at one another and the cat then fled when approached by
observers. On the second occasion a characteristic loris fear cry was
heard. Attracted by this cry, the observers saw a loris being chased
up a tree by a cat, which was also uttering a vocalization. When disturbed, the cat once again fled. The only other case of possible predation threat was when a female loris uttered a fear cry but no
predator was seen. Such a cry was never uttered in response to
another loris, even during the most aggressive fights.

Slender lorises were confronted with potentially dangerous
snakes at least three times. An adult female comfortably fed in a
tree with a pit viper, a snake that the field assistant insisted could
kill a man. Two young lorises remained parked in a tree with a vine
snake for 3.5 hours without showing any signs of fear. An adult
male loris also ignored this same type of snake as he walked over it
as if it were a support. King cobras were also present in the area, but
were never seen near a loris. However, an anecdotal report of
slender loris behavior in Sri Lanka records an attack on a loris from
a king cobra (Still 1905).

Infant lorises are carried by the mother, clinging to her fur for
the first 3 weeks of life but are then parked on exposed locations at
the end of narrow branches where they seem vulnerable to owls. A

total of 917 owlet vocalizations were recorded during 132 hours of observation. Average rates of vocalizations gradually increased and then decreased between dusk and dawn. Owlets also increased their calling frequency in the light phase of the lunar cycle. This may indicate more intense hunting levels, as several species of owl are known to hunt more on brighter nights (Clark 1983). It was expected that variation in loris behavior would correlate to a pattern similar to that of owlet vocalizations and activity. Behaviors such as maximizing tree cover were predicted to increase as the threat of predation increased.

The change in the frequency of owlet vocalizations from dusk to dawn did not correlate with the use of tree cover in either infant or adult age classes. Ordinarily, loris behavior did not change after an owlet called from nearby, regardless of degree of tree cover. On the contrary, in several instances the loris moved to a more exposed position, thus increasing the risk of detection by potential avian predators (Buzzell 1999, Nekaris and Buzzell 2000). There was, however, a difference in activity between the infants and adults on light (moonlit) nights. Only infants altered their activity and maximized tree cover when the moon was up. Based on these observations, *Athene brahma* does not appear to be a threat to adult slender lorises. This is expected as adults fall outside the size range of their prey. However, the infant's increased use of tree cover on light nights may serve to reduce risk of detection by predators in the especially vulnerable age class.

Galagos
Only 12 encounters between galagos and possible predators were observed during all-night follows and eight interactions were witnessed at other times. Ten cases involved genets, two involved domestic cats and two involved jackals. When these carnivores were on the ground they almost invariably provoked alarm calls (whistles and yaps). These were generally given by lone galagos but the calls sometimes induced a communal mobbing response by attracting two or three others, which called frantically for up to 30 minutes while jumping in the vicinity of the carnivore. This resulted in the carnivore moving away. Genets were encountered by galagos in a tree on four occasions but produced no noticeable response except for one genet sitting on a bird's nest. A female with young infants nearby directed a barrage of calling at this genet for 20 minutes, but it stayed firmly in place. One revealing example was seen when a genet moved from the ground into a tree after it had

been spotted by a galago. As soon as the carnivore entered the tree, the galago stopped calling and took no further notice, suggesting that only genets on the ground are perceived as a threat, except in the case of the mother with infants.

Galagos showed a nervous reaction to swooping owls on four occasions, including the shadow of a flying owl cast on the ground by moonlight, but this did not result in calls or other signs of agitation. They simply jumped to safety. Perching eagle owls did produce a mobbing response on two occasions, with persistent calls for 30 minutes – once from a single animal and once from three individuals. Lesser galagos were not seen near snakes at night but a similar mobbing response towards snakes is reported for other galago species (T. Butynski, personal communication; personal observation). When an unhabituated galago is first approached by an observer it will stare for around 20 minutes before settling. If the approach is too rapid, then it will retreat to the top of the tallest available tree, sometimes giving brief alarm calls, then remain silent and immobile until the danger passes. All other species in the study area (cattle, antelope and other herbivores) were ignored.

Infant galagos, like infant lorises, are especially liable to predation since they are often parked for most of the night on their own. Twins are common and are carried one at a time in the mother's mouth, never on the fur. She repeatedly moves them from one nest, or tree hollow, to another prior to 10 days of age when they are placed in trees nearby. The mother continues to return to them during the night to suckle and she carries them from place to place, a behavior that is also common in captivity. As the infants mature and are able to move around on their own, the mother ceases to move them at night and may stay away throughout the night, always returning just before dawn to lead them to a safe sleeping place.

Avoidance of nocturnal threats

Lorises

Though slender lorises are cryptically colored and have quadrupedal locomotion, adults are far from cryptic at night. They are relatively gregarious when compared to many other nocturnal prosimians. Interactions in the vicinity of sleeping sites were friendly and often lasted up to 1 hour past dusk, with animals foraging together and grooming one another. Mysore slender lorises spent almost a quarter of their active time in association with other

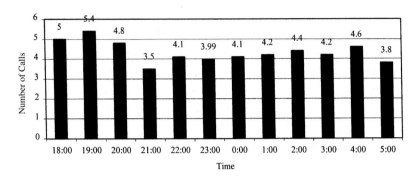

animals throughout the night. Up to six animals were observed within 20 m of each other with no negative reactions and, in fact, 18% of all positive social bouts occurred among three or more animals. They also produced loud calls throughout the night. The principal call heard was the omnipresent whistle, one of five calls made by this subspecies. The whistle consisted of between 0.5 and 6 syllables and could be heard from well over 100 m.

The whistle was heard in many contexts but not in relation to predation. Both females and males made this call in a territorial context to same sex conspecifics at the borders of their home range. Females made this call to males who were attempting to groom them. The males eventually did groom the females, suggesting that the call may have indicated 'agitation' by the female, but not 'territoriality.' Several times a single animal was seen to utter the call when no other animals were in sight, suggesting use of the call as a spacing device similar to that seen in orang-utans (Galdikas 1983, Mitani 1985). Whistles gave an indication of the general direction of the caller, but it proved to be very difficult to pinpoint the location of the animal unless it made several whistles in a row. Only once did the whistle serve as a contact call, when two animals uttered the call and then came together amicably.

Lorises gave these loud whistles at a rate from 3.5 to 5.4 times per hour (Fig. 2.2), with a minimum of 0 and a maximum of 27. The mean for all times of the night was 4.3 whistles per hour. There was no relationship between calling frequency and time (analysis of variance (ANOVA) $p < 0.05$). However, there was a difference between full and new moon. They called more during the new moon than the full moon (t-test, $p < 0.009$). This pattern also applies when dark phases of the night are compared with moonlit phases (ANOVA, sum of squares $= 594.83$, df$=5$, mean square $= 118.97$, F$=6.35$, $p \leq 0.001$) – i.e., they call less frequently when it is light.

Foraging and traveling took place both in the trees and, less often, on the ground. Ground use was minimal, and was only used as a last resort to cross between discontinuous substrates. Lorises moved on the ground for up to 20 m, but did so only after repeated checking for potential dangers and hesitant movements before risking rapid transit to the nearest safe support. However, despite the fact that the greatest potential threat, the wild cat, was terrestrial, lorises used the ground during both dark and light moon phases ($\chi^2 = 1.98$, df $= 2$, $p < 0.37$, $n = 2403$). Other forms of activity, including resting and traveling (movement not associated with feeding), also showed no relation to the moonlight.

Galagos

In contrast to slender lorises, lesser galagos are less continuously vocal and relatively cryptic, although they do come together at night and, on rare occasions, may spend several hours grooming and playing together in groups of two or three. An analysis of 51 all-night follows showed that the focal animals were engaged in social interactions during 14% of the instantaneous samples and were within sight of another galago (<30 m) on a further 7% of samples. Despite having a greater range of calls (20–25 calls in total, Bearder 1999), there were no calls equivalent in frequency to the loris whistle. Barks were made occasionally by both males and females in a territorial or gathering context, and alarm calls occurred sporadically during social encounters and when aroused by dangers (mean, 2 bouts of loud calls per hour, $n = 430$ h).

The state of arousal of individuals gives a useful clue to the seriousness of the disturbance. An undisturbed galago moving quietly through the trees will go to the ground in order to feed or cross an open space with the minimum of visual and auditory checking of the environment. An agitated or nervous animal, in contrast, will often give a variety of alarm calls (including discrete, graded and mixed-unit sequences) and will not descend to the ground without a much more elaborate ritual of trial runs and repeated vigilance.

Given that genets capture galagos when there is no moon they might be expected to seek cover at these times. Radio tracking facilitated precise measurement of their movements. The mean travel distance per hour for all follows was 144 m. However, when the travel path of each individual is divided into periods of twilight, compared to moonlight or no moon, it becomes clear that travel distance is strongly related to levels of ambient light. Adult male galagos moved significantly further each hour during periods of moonlight (158 m) than during periods without a moon (82 m; see

Table 2.1. *The effect of ambient light on speed and range of travel in* Galago moholi *(excluding migrations)*

	Travel speed (m h^{-1})			Night range length (m)		
	No moon	Moon	Twilight	New moon	First/last quarter	Full moon
Adult males						
Mean	82	158	237	327	387	463
Range	47–168	65–240	164–377	150–654	195–640	307–638
S.E.	11.7	16.6	18.2	54.5	47.3	28.3
n	(10)	(11)	(13)	(10)	(11)	(13)
Total radio tracking hours	(103)	(139)	(69)			
Adult females						
Mean	78	109	192	250	247	324
Range	27–140	45–180	83–290	183–418	214–280	192–455
S.E.	11.0	13.4	19.4	34.8	33.0	40.4
n	(11)	(8)	(11)	(7)	(2)	(6)
Total radio tracking hours	(68)	(64)	(33)			

Mann–Whitney U test

Males: No moon vs. moon $U = 14$; $p < 0.01$ New moon vs. full moon $U = 30$;
  Moon vs. twilight $U = 25$; $p < 0.01$ $p < 0.025$

Females: No moon vs. moon $U = 23$; $p < 0.05$ New moon vs. full moon $U = 11$;
  Moon vs. twilight $U = 13$; $p < 0.01$ $p = 0.090$

Table 2.1). Their speed of travel was even greater during twilight (237 m h^{-1}). Finally, judging by measurements of night range lengths, an increase in ambient light also contributes towards wider ranging movements, not merely faster movements within a small area. The situation is similar although less marked in the case of females.

Predation pressure and foraging behavior

Lorises

Lorises increased both their foraging and travel between trees during the light moon, and significantly decreased this behavior during the dark phases of the moon ($\chi^2 = 13.33$, df $= 5$, $p \leq 0.02$, $n = 2403$). Because other behaviors showed no significant difference between light and dark moon phases, it is possible that lorises could see prey better during the light moon and hunted more during this phase. The staple of the diet of lorises was ants and termites, which are widely distributed over bushes and trees and are easy to catch (Nekaris 1999). During the light moon, lorises significantly decreased consumption of ants; however, rate of consumption of other prey items remained the same ($x^2 = 27.73$, df = 14, $p < 0.01$, $n = 753$). Perhaps lorises increase consumption of this food item in the dark moon, as it is more reliable and easier to locate in less light.

Even though foraging behavior increased during the light moon, hunting success did not improve. Lorises were almost exclusively faunivorous, and it was possible to calculate the rate of insects they ate per hour. The average number of prey items caught per hour during the dark moon was 3.76 ± 3.92 ($n = 231$), whereas the average number of prey items caught per hour during the light moon was 3.65 ± 3.64 ($n = 147$). Foraging success remained the same regardless of moon phase (ANOVA, sum of squares = 5.406, df = 1, mean square = 5.406, F = 0.350, $p \leq 0.554$).

Even if foraging success did not change, lorises might be expected to feed in locations less vulnerable to potential predators. Lorises fed mostly on fine branches both in the middle area of trees and on terminal branches, and only came to the ground for food on 14 occasions (Nekaris and Rasmussen 2001). Lorises ate their prey *in situ*; when a prey item was captured, it was eaten immediately. Therefore, during the more vulnerable moon-phase, they might be expected to eat in areas offering more protection from potential predators, such as the middle branches of trees. No significant relationship was found between phase of the moon and feeding location ($x^2 = 4.23$, df = 4, $p \leq 0.38$, $n = 1240$).

Galagos

Foraging in lesser galagos was strongly associated with a single tree species, *Acacia karoo*. During the all-night observations, individuals spent 53.4% of the observation intervals in *A. karoo*, which represented only 25% of the tree species measured by sampling vegetation at random. *A. karoo* provided a year-round source of insects and/or gum, the sole food items consumed by galagos. Figure 2.3 illustrates the main tree species recorded in preferred sleeping and feeding areas compared to an average vegetation plot in the study area. *A. karoo* represents 66.7% of the large trees in the feeding plot. Furthermore, galagos were recorded in *A. karoo* on

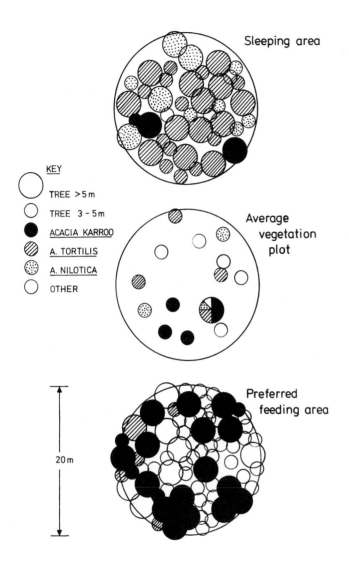

Sleeping area

KEY

○ TREE >5 m

○ TREE 3 - 5 m

● ACACIA KARROO

◨ A. TORTILIS

⬚ A. NILOTICA

○ OTHER

Average vegetation plot

20 m

Preferred feeding area

Fig. 2.3. Diagram showing tree density and species used by *Galago moholi* in a sleeping area (*above*) and a preferred feeding area (*below*) in comparison with an average vegetation plot (*center*).

67.3% of all records of 'foraging' and 'feeding.' There were no significant differences between males and females ($\chi^2=0.524$, df$=1$, $p\leq1.0$, $n=1905$).

Ambient light had an important influence on the proportion of A. *karoo* visited. During 'light' periods of the night, both males and females were recorded in this tree species less frequently (twilight, 45.2%: moonlight, 50.8%) than when there was no moon (61.6%: $\chi^2=15.53$, df$=2$, $p\leq0.003$, $n=4686$). A separate analysis of instantaneous samples when the animals were seen foraging showed that relatively little time was spent foraging during periods of twilight (21.7%) than either moonlight (34.3%) or no moon (34.1%) ($\chi^2=22.263$, df$=2$, $p\leq0.0001$, $n=4012$).

Fig. 2.4. (a) Variations in the path length and night range length for a single male *Galago moholi* (Q2) during four phases of the moon (3 February, 11 February, 28 February, 20 March, 1977). The northern boundary of the study area is included for reference (heavy lines).

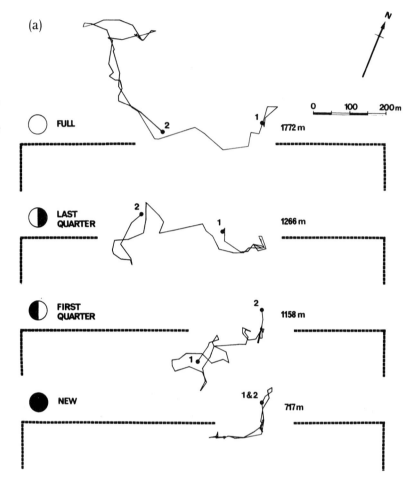

The influence of moonlight is clarified by the case of a well-studied male (Q2), whose nightly movements were recorded at four phases of the moon during one summer (Fig. 2.4a). At the time of the new moon (no visible moon), Q2 traveled rapidly during twilight but spent the rest of the night in a few trees in a dense part of his range where there was a good supply of food (path length = 717 m; night range length = 225 m). On another night when the moon was waxing, he traveled extensively during the first half of the night but retreated to these same trees once the moon set (path length = 1158 m; night range length = 300 m). Exactly the opposite occurred when the moon was visible for the second half of the night (path length = 1266 m; night range length = 425 m) and he traveled throughout the night when the moon was full (path length = 1772 m; night range length = 575 m).

The general pattern of traveling during periods of moonlight

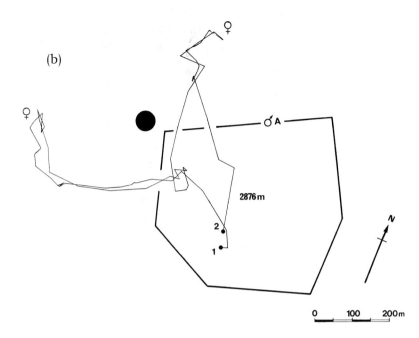

Fig. 2.4. (*cont.*). (b) An example of extensive mating season movements of a male which brought him into contact with two unfamiliar estrus females at the time of a new moon. Heavy lines show the area normally occupied outside the mating season.

and retreating to a safe feeding area when the moon is absent is not, however, universal. During a brief mating season, for example, one male covered nearly 3 km in one night when there was no moon, visiting estrus females far outside his normal home range (Fig. 2.4b).

Discussion and conclusions

It is important to appreciate that lorises and galagos, like other primates, are visually oriented animals. Contrary to popular accounts, they do not follow scent trails through the trees. Their habit of spreading urine on the palms of the hands and soles of the feet (urine washing) is primarily related to enhancing their grip (Welker 1974, Bearder 1974, Charles-Dominique 1977). They also have an excellent sense of smell, but this is more important in a social context and for locating food rather than for navigation. Sensitive hearing is used to detect prey species and predators as well as to communicate by sound. Here we summarize the main differences between the two study species in their reactions to predators and how the risk of predation may influence their foraging. We also discuss the extent to which their behavior has been shaped by both local and evolutionary influences and emphasize the need for further study.

The Mysore slender loris and the Southern African lesser galago appear to be remarkably similar in their lifestyles, despite the fact that one is unable to jump and the other is among the most specialized of the leaping galagos. This chapter provides clear evidence of different predation pressures on the two species. Infant lorises did not react predictably to the calls of spotted owlets but they did seek cover on moonlit nights, which coincides with the time that spotted owlets are most active. Adult lorises did not seek cover and were more active during moonlight hours. Their foraging activity increased with moonlight, yet their hunting success remained the same. It appears that they capture more active prey when it is relatively light and resort to a higher proportion of slow moving species when there is no moon. Lorises were invariably reluctant to travel on the ground at any time, indicating a fear of terrestrial hunting jungle cats, but they used the ground evenly, regardless of moonlight, and they called significantly more often during dark phases. These observations suggest that they are not particularly vulnerable in the dark, in sharp contrast to the situation with lesser galagos.

Galagos are evidently most susceptible to ground predators (genets) that hunt when it is dark and, particularly, when it is also windy. Thus, although they were capable of long distance travel when there was no moon (e.g., during the mating season), they preferred to remain in the safety of tree clumps (usually *A. karoo*) where they could find food without having to cross the ground. This effect was most marked in males, which had larger ranges that overlapped two or more female ranges. Infant galagos may be more vulnerable than lorises, since they cannot travel with the mother on her fur when they are very young. Instead, the mother frequently moves them between nests and parking places until they are old enough to jump to safety. Nevertheless, it is at this stage that most infants are lost.

Both species were sometimes seen in small groups during the night but there is little to suggest that this is a response to predation pressure, except to the extent that galagos may be attracted by the mobbing response of other galagos when a predator has been spotted. This response to danger, and the reaction to conspecifics at night, is another major difference between the two species. Under these circumstances, galagos make themselves conspicuous through a large repertoire of loud calls, which can continue for 30 minutes or more. Theories on the survival value of such mobbing behavior are that it helps to alert close relatives to danger (kin selection) and that it benefits the caller by informing the predator that it has been detected, after which further hunting is futile (Cheney and Wrangham 1987, Harvey and Greenwood 1978). Since galagos usually encounter predators when alone, their calling most probably informs the predator that it is not worth continuing. This is supported by observations of carnivores moving away once detected. However, calls are also given by females with infants, which suggests that kin selection is also involved.

Most textbook summaries of the behavior of lorises and galagos report the latter to be more active, gregarious and noisier. Indeed, the West African pottos and angwantibos tend to spend more time alone, and do not produce a loud call (Charles-Dominique 1977; E. Pimley, personal communication). The slow loris (*Nycticebus coucang*), another cryptic form, is reported to emit a loud call, but does so infrequently (F. Wiens, personal communication). From the perspective of an animal that cannot leap to safety, an alarm call would seem to be more dangerous than beneficial. Galagos, on the other hand, can call with less fear of harm as they can escape swiftly. Infant galagos, like lorises, generally remain silent and, when they do call in distress, or when following the mother, their high-pitched clicking calls are not easy to locate.

Despite the absence of mobbing calls, slender lorises could be heard making loud whistles throughout the night. Very few potential predators occurred in the study area and it is possible that the less cryptic behaviors in this particular population are a recent artifact due to elimination of most predators. During a preliminary survey of various slender loris habitats in South India, the loud whistle was heard even when few or no animals were located, though the actual rates of calling were not quantified (Nekaris 1997). In particular, the whistle was common in Mundanthurai Tiger Reserve where population densities are known to be higher, and a full complement of predators still exists (Kar Gupta 1995; personal observation). Though the whistle of the loris was audible from 200–300 m, the call may not be so easy for a predator to pinpoint. Based on observations of the slow loris in captivity, Daschbach *et al.* (1981) claimed that detection is actually difficult because calls are widely spaced, have a narrow frequency range, and are long in duration.

Even if predators have been eliminated relatively recently from AIFD, the evolutionary effects of past predation events should still be reflected in loris behavior. Predation events may be infrequent over the short term but may be more significant over time and certainly over the history of a species. This long-term significance is reflected in antipredator behavior in the absence of a present day predator (Anderson 1986). Evidence for this phenomenon has been demonstrated in some lemurs, which have a wide repertoire of loud calls for various aerial predators even though such predators have long been extinct on Madagascar (Goodman *et al.* 1993).

Crypsis in *Loris tardigradus* has largely been described so far as a strategy to reduce chances of an encounter with predators (Sussman 1999) but such behavior will not protect an animal from an imminent attack. The slender loris lacks defense morphology such as the scapular shield of *Perodicticus potto* (Rasmussen 1986). The skin in the nuchal area, however, is double the thickness of the skin on the rest of the body. Schulze and Meier (1995) suggested that this thickness could be a last-minute protective defense. Mysore slender lorises did indeed show cryptic behaviors, such as noiseless movement and freezing when startled. However, studies in other habitats and in areas with a larger complement of predators are necessary to evaluate how much of their behavior is influenced by this factor.

In conclusion, long-term observations made on two nocturnal primates show that it is possible to collect useful data on predation risk and its possible influence on foraging. Our studies point to the value of recording *ad hoc* observations that may eventually add up to a meaningful pattern. They show that, if anything, adult lesser galagos are more cryptic than slender lorises, contrary to popular impressions. At least some species of nocturnal primates lend themselves to detailed study, often because of the fact that they do not feel threatened by human observers at night and become fully habituated to a quiet approach within 30 minutes – even if some of them are afraid of the dark.

Acknowledgments

Our sincere thanks go to the many people who provided advice, support and assistance, particularly Mewa Singh, Sindhu Radhakrishna, Tab Rasmussen, Bob Sussman, Charles Southwick, Bob Martin, Catherine Bearder, Richard and Barbara Galpin and family, Sam Pullen, Dan Freeman, Todd Olson and Gerald Doyle. We gratefully acknowledge the support and patience of Lynne Miller and her editorial team.

REFERENCES

Ali, S. (1964). *The Book of Indian Birds*, 7th Edition. Bombay: Bombay Natural History Society.

Ali, S. (1969). *Birds of Kerala*, 2nd Edition. Oxford: Oxford University Press.

Altmann, J. (1974). Observational study of behavior: sampling methods. *Behavior* **49**: 227–65.

Anderson, C.M. (1986). Predation and primate evolution. *Primates* **27**: 15–39.

Andrews, P., and Evans, E.M. (1979). Small mammal bone accumulations produced by mammalian carnivores. *Palaeobiology* **9**: 289–307.

Bearder, S.K. (1969). Territorial and intergroup behavior of the lesser bushbaby, *Galago senegalensis moholi* (A. Smith), in semi-natural conditions and in the field. MSc dissertation, University of the Witwatersrand, Johannesburg.

Bearder, S.K. (1974). Aspects of the ecology and behavior of the thick-tailed bushbaby *Galago crassicaudatus*. PhD thesis, University of the Witwatersrand, Johannesburg.

Bearder, S.K. (1987). Lorises, bushbabies and tarsiers: diverse societies in solitary foragers. In: B.B. Smuts, D.L. Cheney, R.M. Seyfarth, R.W. Wrangham and T.T. Struhsaker, eds., *Primate Societies*, Chicago: University of Chicago Press, pp. 11–24.

Bearder, S.K. (1999). Physical and social diversity among nocturnal primates: a new view based on long term research. *Primates* **40**: 267–82.

Bearder, S.K., and Martin R.D. (1980). The social organization of a nocturnal primate revealed by radio tracking. In: C.J. Amlaner Jr., and D.W. Macdonald, eds., *A Handbook on Biotelemetry and Radio Tracking*, Oxford: Pergamon Press, pp. 633–48.

Buzzell, C.A. (1999). Field observations of antipredator behavior in the slender loris (*Loris tardigradus lydekkerianus*). Undergraduate Senior Honors Thesis. Washington University, St. Louis.

Charles-Dominique, P. (1977). *Ecology and Behavior of Nocturnal Primates*. London: Duckworth.

Charles-Dominique, P., and Bearder, S.K. (1979). Field studies of lorisid behavior: the lorisids of Gabon; the galagines of South Africa. In: G.A. Doyle and R.D. Martin, eds., *The Study of Prosimian Behavior*, New York: Academic Press, pp. 567–629

Clark, J.A. (1983). Moonlight's influence on predatory prey interactions between short-eared owls (*Asio flammeus*) and deermice (*Peromyscus manicularus*). *Behavioral Ecology and Sociobiology* **13**: 205–9.

Cheney, D.L., and Wrangham, R.W. (1987). Predation. In: B.B. Smuts, D.L. Cheney, R.M. Seyfarth, R.W. Wrangham and T.T. Struhsaker, eds., *Primate Societies*, Chicago: University of Chicago Press, pp. 227–39.

Curtis, D.J., and Rasmussen, M.A. (2001). Cathemerality in lemurs. In: *Proceedings of the Centenary Congress of the Anthropological Institute and Museum, Zurich 1899–1999: Primatology and Anthropology into the third millennium* (in press).

Daschbach, N.J, Schein, M.W., and Haines, D.E. (1981). Vocalizations of the slow loris, *Nycticebus coucang* (Primates, Lorisidae). *International Journal of Primatology* **2**: 71–80.

Dunbar, R.I.M. (1977). *Primate Social Systems: Studies in Behavioural Adaptation.* London: Croom Helm.

Galdikas, B.W.F. (1983). The orang-utan long call and snag crashing at Tanjun Putting Reserve. *Primates* **24**: 371–84.

Goodman, S.M., O'Connor, S., and Langrand, O. (1993). A review of predation on lemurs: Implications for the evolution of social behavior in small nocturnal primates. In: P.M. Kappeler and J.U. Ganzhorn, eds., *Lemur Social Systems and Their Ecological Basis,* New York: Plenum Press, pp. 51–66.

Harvey, P.H., and Greenwood, P.J. (1978). Anti-predator defense strategies: Some evolutionary problems. In: J.R. Krebs and N.B. Davies, eds., *Behavioural Ecology: An Evolutionary Approach,* Sunderland, MA: Sinauer Associates, pp. 129–151.

Kar Gupta, K. (1995). Slender loris, *Loris tardigradus,* distribution and habitat use in Kalakad-Mundanthurai Tiger Reserve, India. *Folia Primatologica* **69 (suppl.)**: 401–2.

Mikkola, H. (1983). *Owls of Europe.* Vermillion, SD: Buteo Books.

Mitani, J.C. (1985). Sexual selection and adult male orang-utan long calls. *Animal Behavior* **33**: 272–83.

Nekaris, K.A.I. (1997). A preliminary survey of the slender loris (*Loris tardigradus*) in South India. *American Journal of Physical Anthropology* **24 (suppl.)**: 176–7.

Nekaris, K.A.I. (1999). The diet of the slender loris (*Loris tardigradus lydekkerianus*) in Dindigul, South India. *American Journal of Physical Anthropology* **28 (suppl.)**: 209.

Nekaris, K.A.I. (2000a). The socioecology of the Mysore slender loris (*Loris tardigradus lydekkerianus*) in Dindigul, Tamil Nadu, South India. Ph.D. Dissertation, Washington University, St. Louis.

Nekaris, K.A.I. (2000b). The activity budget of the slender loris (*Loris tardigradus lydekkerianus*) in Dindigul, South India. *American Journal of Physical Anthropology* **30 (suppl.)**: 235–6.

Nekaris, K.A.I., and Buzzell, C.A. (2000). How cryptic is the slender loris? *American Journal of Primatology Supplement* **51**: 76–7.

Nekaris, K.A.I and Rasmussen, D.T. (2001). The bug-eyed slender loris: insect predation and its implications for primate origins. *American Journal of Physical Anthropology* **32 (suppl.)**: 112.

Overdorff, D.J., and Rasmussen, M.A. (1995). Determinants of nightime activity in 'diurnal' lemuroid primates. In: L. Alterman, G.A. Doyle and M.K. Izard, eds., *Creatures of the Dark: The Nocturnal Prosimians,* New York: Plenum Publishing, pp. 61–74.

Rasmussen, D.T. (1986). Life history and behavior of slow lorises and slender lorises. PhD thesis, Duke University, Durham, NC.

Rasmussen, M.A. (1999). Ecological influences on activity cycle in two cathemeral primates, *Eulemur mongoz* (mongoose lemur) and *Eulemur fulvus fulvus* (common brown lemur). PhD thesis, Duke University, Durham, NC.

Schulze, H., and Meier, B. (1995). Behavior of captive *Loris tardigradus nordicus*: A qualitative description including some information about morphological bases of behavior. In: L. Alterman, G.A. Doyle and M.K. Izard, eds., *Creatures of the Dark: The Nocturnal Prosimians*, New York: Plenum Publishing, pp. 221–250.

Still, J. (1905). On the loris in captivity. *Spolia Zeylanica* **3**: 155–157.

Sussman, R.W. (1999). *Primate Ecology and Social Structure* Vol. I. Pearson Custom Publishing: MA.

Tattersall, I. (1987). Cathemeral activity in primates: A definition. *Folia Primatolologica* **49**: 200–2.

Welker, C. (1974). Verhaltensphysiologische Untersuchungen an *Galago crassicaudatus* E. Geoffroy, 1812 (Lorisiformes: Galagidae) in Gefangenschaft. Inaugural Doctoral Dissertation, Justus Liebig-Universitat, Giessen.

3 • Predation sensitive foraging in captive tamarins

MARK J. PRESCOTT & HANNAH M.
BUCHANAN-SMITH

Introduction

Tamarins of the genus *Saguinus* are small bodied (250–650 g: Hershkovitz 1977, Napier and Napier 1967, Snowdon and Soini 1988), Neotropical, diurnal, arboreal frugivore–insectivores, some of which form stable mixed-species troops (saddle-backed tamarin (*S. fuscicollis*) with either moustached tamarin (*S. mystax*), red-bellied tamarin (*S. labiatus*) or emperor tamarin (*S. imperator*): see Heymann and Buchanan-Smith 2000 for a review). They are vulnerable to a wide range of predatory birds, reptiles and mammals (Buchanan-Smith 1990, Dawson 1979, Emmons 1987, Hershkovitz 1977, Heymann 1987, 1990, Izawa 1978, Neyman 1978, Terborgh 1983), and it has been suggested that, due to their small size, they have higher rates of predation relative to large-bodied primates (Caine 1993, Cheney and Wrangham 1987, Terborgh 1990).

Given such high predation risk, and given that they typically spend around 60% of their daily activity period foraging for their food (Soini 1987, Terborgh 1983), it is expected that antipredatory considerations may impose strong constraints on tamarin foraging behavior. The antipredatory behavior of callitrichines suggests that early detection and consequent avoidance of predators is the most effective mechanism to decrease predation. For example, in response to aerial predation, tamarins appear to rely on crypsis (i.e., remaining immobile, hiding beneath foliage), retreating from the periphery of a tree (where many of their fruits are found) to the tree trunk or, most dramatically, free-falling to the ground (e.g., Caine 1987, Dawson 1979, Heymann 1990, Peres 1991). Heymann (1990) observed a mixed-species troop of saddle-backed tamarins and moustached tamarins to remain immobile for almost 4 hours following an observed attack by the ornate hawk-eagle (*Spizaetus ornatus*).

Observations such as these suggest that, for tamarins, vigilance to allow early detection of predators followed by crypsis is more important than any other antipredatory strategy, such as fleeing in concert or mobbing. Yet such a strategy is incompatible with foraging. In this regard, in a comparison of activity budgets before and after alarm calling, Peres (1993) found that, following an alarm call, saddle-backed tamarins and moustached tamarins remained quiet in the understory and spent significantly less time feeding and foraging. It is possible that tamarins cut short their foraging time in order to travel to safe sleeping sites before dusk (Caine 1987, Moynihan 1970). In addition, in captivity, preferences have been demonstrated for selecting the nearest neighboring feeding patch in tamarins (Prescott 1999 for saddle-backed tamarins and red-bellied tamarins) and marmosets (MacDonald et al. 1994 for the common marmoset (*Callithrix jacchus*)), which may reduce exposure to predation.

Furthermore, when food is found, even if only in small amounts, food calls are given almost invariably. These are thought to recruit troop-mates to the vicinity of the caller in order to maintain troop-mediated vigilance (Caine et al. 1995 for red-bellied tamarins). This near-invariant response seems to indicate that individual tamarins may be forced to balance exclusive exploitation of any small monopolizeable resource they have discovered against increased feeding competition due to the necessity for maintaining social contact and cohesion for cooperative vigilance (Prescott 1999). In tamarin troops, where troop cohesion and strong mutual interdependence of troop members is thought to have fundamental survival value (Caine et al. 1995), the need to follow one's troop-mates or troop-leaders, or simply to keep one's troop-mates in auditory or visual contact, is likely to constrain the movement (and hence foraging) of individual animals about their home range quite considerably. The foraging behavior of individuals in mixed-species troops may be particularly constrained in this regard, given that, in order to keep the association intact, at least one species may have to deviate from its optimal foraging pattern in order to follow the other species from which its diet differs.

We believe that, as prey animals, tamarins face a daily conflict because behavioral options leading to maximally efficient food intake may leave them vulnerable to predators. Given this thinking, we decided to investigate the foraging height preferences of saddle-backed tamarins and red-bellied tamarins in captive single-species troops to see if foraging behavior is altered in relation to perceived threat of predation. Captive tamarins exhibit many of the behavioral patterns typical of their wild counterparts, lending support to the generalization of results to other troops both in the wild and in captivity (see Buchanan-Smith and Hardie 1997, Hardie *et al.* 1993, Prescott 1999).

Although the most important predators of tamarins are medium to large-sized diurnal raptors (Goldizen 1987, Heymann 1990, Peres 1993, Sussman and Kinzey 1984, Terborgh 1983), carnivorous terrestrial and scansorial mammals, such as small to medium-sized felids and mustelids, pose a threat to callitrichines (Emmons 1987, Stafford and Ferreira 1995). Anecdotal reports of attempted predation on tamarins by terrestrial/scansorial predators include attacks by ocelots (*Felis pardalis*: Heymann 1990) and tayras (*Eira barbara*: Buchanan-Smith 1990, Jansen personal communication to Galef *et al.* 1976, Ramirez 1989, Smith personal communication to Moynihan 1970). Tayras are very agile in the trees (Leopold 1959), but ocelots, although good climbers, do most of their hunting on the ground (Eisenberg and Redford 1999). Wild tamarins are extremely cautious of descending to the forest floor (Buchanan-Smith 1991; Heymann personal communication for moustached tamarins; Prescott personal observation for red-bellied tamarins). Wild saddle-backed tamarins and red-bellied tamarins are more vigilant lower in the forest (<8 m, 25% of samples) compared to higher in the forest (>8 m, 14% of samples) (Buchanan-Smith unpublished data) although these data may be confounded by differential visibility (see Ferrari and Rylands 1994; Treves, Chapter 14). Since there is evidence in many species of primate that vigilance increases when predation risk increases (Caine 1984, Chapman and Chapman 1996, Cheney and Seyfarth 1990, Hauser and Wrangham 1990, Macedonia and Yount 1991, Zuberbühler *et al.* 1997), then it is a reasonable assumption that when vigilance is greater, predation risk is greater. Given this assumed threat of attack from the forest floor, one would expect the perceived risk of predation to be greatest near to the ground and consequently a foraging height at a greater distance away from the ground to be preferred over one closer to the ground. This was investigated in Experiment 1.

Experiment 1

Foraging boxes containing one of two different quantities of food were presented at one of two different heights to single-species troops of saddle-backed tamarins and red-bellied tamarins to determine if the species exhibited foraging height preferences.

Method
Study animals, housing and husbandry
The study was conducted at Belfast Zoological Gardens, Northern Ireland, where the study animals were housed in enclosures off-exhibit to the public to minimize disturbance. Six troops of saddle-backed tamarins and six troops of red-bellied tamarins were tested. Each troop consisted of an adult (>18 months of age) male and female pair together with any offspring, of which only the adult pair was tested. The monkeys' ages were similar across species (the mean age for saddle-backed tamarins was 4 years 7 months and for red-bellied tamarins was 4 years and 5 months). All individuals were captive born.

Each troop occupied their own standard captive indoor/outdoor enclosure (2 m (ht.) × 1.75 m × 1.5 m/1.95 m (ht.) × 1.55 m × 3.5 m). Further details of housing and of husbandry are given in Prescott and Buchanan-Smith (1999). Troops were tested in a large outdoor enclosure (approx. 210 m³, with a height of 3.7 m at one end and 5 m at the other). All monkeys had previously had access to the enclosure, which was well furnished with a network of fixed and nonfixed dead branches and natural vegetation. The monkeys were moved to an adjacent indoor/outdoor enclosure, where they spent at least 48 hours, prior to access being given to the large testing enclosure.

Experimental design
Each troop received four separate test trials, during which a foraging box containing one of two possible quantities of food was presented to them at one of two possible heights. The foraging box was a semi-transparent polypropylene container measuring approximately 13 cm × 8 cm × 6 cm and patterned with one of two different designs (vertical stripes or filled circles). The designs were used to indicate that the box contained one of two possible quantities of food (mealworms concealed beneath a 3 cm layer of sterile wood shavings). The monkeys were familiar with foraging for food items from the foraging boxes and were trained to associate a particular design with a particular quantity of hidden food. (This was necessary for Experiment 2, which investigated whether preferred foraging height would be traded-off against food quantity. Food quantity was not analyzed in Experiment 1.) Details of the training

procedure are given in Prescott (1999). The designs were assumed to be equally visible and were counterbalanced to eliminate preference effects (i.e., for half the troops, vertical stripes indicated five food items, and filled circles, 20 food items; for the other half, vertical stripes indicated 20 food items and filled circles, five food items). The foraging box was fixed to a large (>10 cm in diameter), horizontal branch at either 50 cm or 150 cm from the ground, according to test condition. Average height use in the enclosure was previously calculated as approximately 125 cm from the ground for both species. The four trials were the four possible combinations of food quantity and height (i.e., five food items at 50 cm; 20 food items at 50 cm; five food items at 150 cm; 20 food items at 150 cm). The order of trial presentation was counterbalanced across troops to control for order effects.

Although predation is not a real threat in their captive environment, both species were extremely vigilant towards terrestrial predators such as feral cats (*Felis silvestris catus*), of which there were nine at the zoo at the time of testing, and other terrestrial animals such as peacocks (*Pavo cristatus*). The frequency of encounters with cats was approximately twice per day, and with peacocks approximately four times per day (Prescott personal observation). Both cats and peacocks were able to walk freely right beside the testing enclosure, to which the monkeys responded with visual fixation, piloerection and alarm calling. However, no such potential threats were present during testing.

Procedure

Test trials were conducted before the monkeys' daily feed to ensure they were motivated to search for the food items. The troop under study was released into the large testing enclosure from an adjoining enclosure after the appropriate quantity of food had been placed in the appropriate foraging box. Recording began the moment the connecting door between this enclosure and the large testing enclosure was opened (from the outside via a wire pulley). Recording continued for a period of 20 minutes, or until all the food items were thought to have been consumed and no monkey had approached within 15 cm of either box for over 5 minutes.

Recording methods and data analysis

All data were recorded by one of the authors (MJP). The sampling strategy was scan sampling (Martin and Bateson 1986). A miniature tape recorder was used to dictate a verbal record of, amongst other

Table 3.1. *Median latencies and ranges (seconds) to obtain a food item from boxes with five and 20 food items at 150 cm and 50 cm from the ground for saddle-backed tamarins and red-bellied tamarins*

Species	Quantity	150 cm	50 cm	Z value
Saddle-backed tamarins	5 items	30.5 (96)	66.0 (72)	$z = -2.04$, **$p < 0.05$**
Saddle-backed tamarins	20 items	38.5 (90)	65.5 (88)	$z = -1.65$, $p > 0.05$
Red-bellied tamarins	5 items	49.5 (101)	93.5 (109)	$z = -2.04$, **$p < 0.05$**
Red-bellied tamarins	20 items	23.5 (29)	100.0 (124)	$z = -2.93$, **$p < 0.01$**

Notes:
P-values refer to Wilcoxon Signed-Ranks test. Bold indicates a significant result.

things, all instances in which an individual approached within 15 cm of the box, obtained a food item, and exited the area within 15 cm of the box ('all-occurrences' recording: Altmann 1974). Data were subsequently transcribed from audio-tape onto record sheets. From the data collected, it was possible to derive latencies (in seconds), from entering the testing enclosure for each individual in each trial to obtain its first food item. It was also possible to calculate for each individual its total duration spent within 15 cm of the box.

Statistical comparisons between heights or condition were made using the Wilcoxon Signed-Ranks test. Comparisons between species were made using the Mann–Whitney U test. Nonparametric statistical tests were used because of sample-size limitations ($n = 12$ for the tables that follow) and deviations from normality. Significance is reported as either alpha <0.05 (significant) or alpha <0.01 (highly significant).

Results
Considering the latency data first, saddle-backed tamarins were significantly faster to obtain a food item from the box with five food items when it was presented at 150 cm from the ground than at 50 cm from the ground (Table 3.1). They exhibited no significant difference in their latency to obtain a food item from the box with 20 food items. Red-bellied tamarins were significantly faster to obtain a food item from the box with five food items and that with 20 food items when these were presented at 150 cm from the ground than at 50 cm from the ground.

Using the same data but rearranging them to show species differences, there was no species difference in the latency to obtain a

Table 3.2. *Species differences in median latencies and ranges (seconds) to obtain a food item from boxes with five and 20 food items at 150 cm and 50 cm from the ground*

Height	Quantity	Saddle-backed tamarins	Red-bellied tamarins	Z value
150 cm	5 items	30.5 (96)	49.5 (101)	$z=-1.33, p>0.05$
150 cm	20 items	38.5 (90)	23.5 (29)	$z=-2.66, \boldsymbol{p<0.01}$
50 cm	5 items	66.0 (72)	93.5 (109)	$z=-2.60, \boldsymbol{p<0.01}$
50 cm	20 items	65.5 (88)	100.0 (124)	$z=-2.16, \boldsymbol{p<0.05}$

Notes:
P-values refer to Mann–Whitney U test. Bold indicates a significant result.

food item from the box with five food items at 150 cm from the ground (Table 3.2). Red-bellied tamarins were found to be significantly faster than saddle-backed tamarins to obtain a food item from the box with 20 food items at 150 cm from the ground. Saddle-backed tamarins were significantly faster than red-bellied tamarins to obtain a food item from the box with five food items and that with 20 food items at 50 cm from the ground.

Now considering the duration data, there was a trend for both species to spend longer within 15 cm of the box when it was presented at 150 cm from the ground than at 50 cm from the ground, regardless of the number of items it contained (Table 3.3). However, this trend was only statistically significant for saddle-backed tamarins when the box contained 20 items and for red-bellied tamarins when the box contained five items.

Using the same data but rearranging them to show species differences, there were no species differences in duration spent within 15 cm of any of the boxes (Table 3.4).

Discussion

Red-bellied tamarins appear to have a preference for the upper box. That is, they showed a reduced latency to this box compared to the lower box, were faster than the saddle-backed tamarins to this box when it contained 20 food items, and spent longer within 15 cm of this box compared to the lower box when it contained five food items. Saddle-backed tamarins show a less-clear preference

Table 3.3 *Median durations and ranges (seconds) spent within 15 cm of the boxes with five and 20 food items at 150 cm and 50 cm from the ground for saddle-backed tamarins and red-bellied tamarins*

Species	Quantity	150 cm	50 cm	Z value
Saddle-backed tamarins	5 items	90.5 (129)	80.0 (143)	$z = -0.31, p > 0.05$
Saddle-backed tamarins	20 items	198.5 (299)	140.0 (190)	$z = -2.20, \mathbf{p < 0.05}$
Red-bellied tamarins	5 items	110.5 (114)	62.0 (141)	$z = -2.69, \mathbf{p < 0.01}$
Red-bellied tamarins	20 items	197.5 (322)	129.0 (189)	$z = -1.57, p > 0.05$

Notes:
P-values refer to Wilcoxon Signed–Ranks test. Bold indicates a significant result.

Table 3.4. *Species differences in median durations and ranges (seconds) spent within 15 cm of the boxes with five and 20 food items at 150 cm and 50 cm from the ground*

Height	Quantity	Saddle-backed tamarins	Red-bellied tamarins	Z value
150 cm	5 items	90.5 (129)	110.5 (114)	$z = -0.87, p > 0.05$
150 cm	20 items	198.5 (299)	197.5 (322)	$z = -0.66, p > 0.05$
50 cm	5 items	80.0 (143)	62.0 (141)	$z = -0.87, p > 0.05$
50 cm	20 items	140.0 (190)	129.0 (189)	$z = -0.03, p > 0.05$

Notes:
P-values refer to Mann–Whitney U test. Bold indicates a significant result.

for the upper box. That is, they were faster to this box compared to the lower box only when it contained five food items and spent longer within 15 cm of this box compared to the lower box only when it contained twenty food items. Saddle-backed tamarins were, however, faster than red-bellied tamarins to the lower box when it contained five food items and when it contained 20 food items.

In investigations of general height use in mixed-species troops in captivity, saddle-backed tamarins and red-bellied tamarins have been found to segregate themselves vertically with saddle-backed tamarins occupying a lower mean height than their congeners (Hardie *et al.* 1993), as is also the case in the wild (Buchanan-Smith 1990, 1999, Hardie 1998, Pook and Pook 1982, Yoneda 1981, 1984). This pattern was not evident in this study. Although both species were found to exhibit foraging height preferences in single-species troops, the preference of both species was to forage at a position high in their enclosure (i.e., they were both generally faster to feed from the upper box than the lower box, and spent longer in proximity to the upper box). This pattern is likely to be due to a general reluctance to descend near to the ground, due to the perceived threat of terrestrial attack, and as such is evidence of predation sensitive foraging in captive tamarins.

However, although both species preferred the upper box, saddle-backed tamarins were faster than red-bellied tamarins to descend to the lower box and obtain a food item from it. This pattern is consistent with data on height preferences in the wild with red-bellied tamarins occupying a higher stratum than their congener and rarely descending to the forest floor, whereas saddle-backed tamarins use all levels down to the forest floor (a consequence of their searching tree trunks for embedded, hidden prey) (Buchanan-Smith 1999, Hardie 1998, Pook and Pook 1982, Yoneda 1981, 1984). Saddle-backed tamarins are thus more likely to descend to the floor, and will do so in the wild to retrieve insects flushed from higher levels by their congeners (Garber 1992, 1993, Norconk 1990, Peres 1991, Terborgh 1983, Yoneda 1981).

Experiment 2

The aim of Experiment 2 was to investigate whether foraging height preferences alter when the monkeys are given a choice between two quantities of food presented simultaneously at different heights (i.e., how do food quantity and predation risk interact?). One would expect both species to be faster to the box containing the greater quantity of food when it is presented high in the enclosure at the preferred height but not when it is presented low at the nonpreferred height. In addition, as saddle-backed tamarins were more willing than red-bellied tamarins to descend low in the enclosure in Experiment 1, we would predict that they would trade preferred foraging height for increased food quantity, but that red-bellied tamarins would not.

Method
Study animals, experimental design, procedure and recording methods
The study animals, experimental design, procedure and recording methods were exactly as in Experiment 1, except that two boxes

Table 3.5. *Median latencies and ranges (seconds) to obtain a food item from boxes presented simultaneously with five and 20 food items at 150 cm and 50 cm from the ground for saddle-backed tamarins and red-bellied tamarins*

Species	Height and Quantity	Latency	Z value
Saddle-backed tamarins	HIGH-20	16.0 (13)	
Saddle-backed tamarins	LOW-5	127.5 (107)	$z = -3.06$, $p < 0.01$
Saddle-backed tamarins	LOW-20	45.5 (106)	
Saddle-backed tamarins	HIGH-5	94.0 (173)	$z = -1.77$, $p > 0.05$
Red-bellied tamarins	HIGH-20	14.0 (11)	
Red-bellied tamarins	LOW-5	115.0 (185)	$z = -3.06$, $p < 0.01$
Red-bellied tamarins	LOW-20	46.5 (73)	
Red-bellied tamarins	HIGH-5	111.0 (79)	$z = -2.51$, $p < 0.01$

Notes:
P-values refer to Wilcoxon Signed–Ranks test. Bold indicates a significant result.

were presented simultaneously in each test trial as opposed to one. Each troop received two test trials, one in which five mealworms were presented at 50 cm from the ground (LOW-5) and 20 at 150 cm from the ground (HIGH-20), and one in which 20 mealworms were presented at 50 cm from the ground (LOW-20) and five at 150 cm from the ground (HIGH-5). Again, the order of trial presentation was counterbalanced across troops to control for order effects.

Data analysis
Data analysis was exactly as in Experiment 1. For brevity, only data for the latency for each monkey to obtain its first food item is presented here.

Results
As we predicted, both saddle-backed tamarins and red-bellied tamarins were significantly faster to obtain a food item from the HIGH-20 box than from the LOW-5 box (Table 3.5). However, contrary to our prediction concerning species differences, there was no significant difference in the latency for saddle-backed tamarins to obtain a food item from the LOW-20 box and the HIGH-5 box. Rather, it was

red-bellied tamarins that were significantly faster to obtain a food item from the LOW-20 box than the HIGH-5 box.

Discussion

If two foraging boxes are presented simultaneously at 150 cm and 50 cm from the ground, and the tamarins have received training to recognize that different designs on the boxes indicate that they contain one of two possible quantities of food, saddle-backed tamarins prefer to feed from the box containing the greater quantity of food items (i.e., the latency to the box with greater quantity of food items is less than that with the lesser quantity of food items) only when it is presented at the preferred height. Red-bellied tamarins prefer this box regardless of the height at which it is presented. So red-bellied tamarins, at least, will trade-off their preferred foraging height for food quantity (their avoidance of the ground is less strong than their attraction to large resources). This finding provides further evidence for predation sensitive foraging in captive red-bellied tamarins. The reason why red-bellied tamarins and not saddle-backed tamarins traded preferred foraging height for food quantity is not clear, but it may reflect fundamental differences between the species in their foraging strategies and risk-taking behavior.

A comparison of foraging behavior in species that form polyspecific associations in the wild is of particular interest as divergent foraging strategies may provide clues to their association patterns, and the advantages that individuals gain and the disadvantages they incur from association. Identical studies have been conducted with mixed-species troops of saddle-backed tamarins and red-bellied tamarins, and these have elucidated how the presence of a congener affects predation sensitive foraging behavior (Prescott and Buchanan-Smith unpublished data).

Acknowledgments

The research was supported by a University of Stirling studentship awarded to Mark J. Prescott. We gratefully acknowledge John Stronge (Zoo Manager) and Mark Challis (Assistant Zoo Manager) at Belfast Zoological Gardens for the opportunity to conduct the research. We are grateful also to the keepers at the zoo for maintaining the study animals. We thank Lynne Miller for inviting us to contribute to the volume and Lynne, Adrian Treves, and Nancy Caine for constructive comments on the manuscript.

REFERENCES

Altmann, J. (1974). Observational study of behaviour: Sampling methods. *Behaviour* **49**: 227–67.

Buchanan-Smith, H.M. (1990). Polyspecific association of two tamarin species, *Saguinus labiatus* and *Saguinus fuscicollis*, in Bolivia. *American Journal of Primatology* **22**: 205–14.

Buchanan-Smith, H.M. (1991). A field study on the red-bellied tamarin, *Saguinus l. labiatus* in Bolivia. *International Journal of Primatology* **12**: 259–76.

Buchanan-Smith, H.M. (1999). Forest utilization and stability of mixed-species groups. *Primates* **40**: 233–47.

Buchanan-Smith, H.M., and Hardie, S.M. (1997). Tamarin mixed-species groups: An evaluation of a combined captive and field approach. *Folia Primatologica* **68**: 272–86.

Caine, N.G. (1984). Visual scanning by tamarins. *Folia Primatologica* **43**: 59–67.

Caine, N.G. (1987). Vigilance, vocalisations and cryptic behaviour at retirement in captive groups of red bellied tamarins (*Saguinus labiatus*). *American Journal of Primatology* **12**: 241–50.

Caine, N.G. (1993). Flexibility and co-operation as unifying themes in *Saguinus* social organization and behaviour: the role of predation pressures. In: *Marmosets and Tamarins: Systematics, Behaviour, and Ecology*, A.B. Rylands, ed., Oxford: Oxford University Press, pp. 200–19.

Caine, N.G., Addington, R.L., and Windfelder, T.L. (1995). Factors affecting the rates of food calls given by red-bellied tamarins. *Animal Behaviour* **50**: 53–60.

Chapman, C.A., and Chapman, L.J. (1996). Mixed-species primate groups in the Kibale Forest: Ecological constraints on association. *International Journal of Primatology* **17**: 31–50.

Cheney, D.L., and Seyfarth, R.M. (1990). *How Monkeys See the World*. Chicago: University of Chicago Press.

Cheney, D.L., and Wrangham, R. (1987). Predation. In: *Primate Societies*, B.B. Smuts, D.L. Cheney, R.M. Seyfarth, R.W. Wrangham and T.T. Struhsaker, eds., Chicago: University of Chicago Press, pp. 227–39.

Dawson, G.A. (1979). The use of time and space by the Panamanian tamarin, *Saguinus geoffroyi*. *Folia Primatologica* **31**: 253–84.

Eisenberg, J.F., and Redford, K.H. (1999). *Mammals of the Neotropics*, Vol. 3: *The Central Neotropics: Ecuador, Peru, Bolivia, Brazil*. Chicago: University of Chicago Press.

Emmons, L.H. (1987). Comparative feeding ecology of felids in a neotropical rainforest. *Behavioural Ecology and Sociobiology* **20**: 271–83.

Ferrari, S.F., and Rylands, A.B. (1994). Activity budgets and differential visibility in field studies of three marmosets (*Callithrix* spp.). *Folia Primatologica* **63**: 78–83.

Galef, B.G., Mittermeier, R.A., and Bailey, R.C. (1976). Predation by tayra (*Eira barbara*). *Journal of Mammology* **54**: 152–4.

Garber, P.A. (1992). Vertical clinging, small body size and the evolution of feeding adaptations in the Callitrichinae. *American Journal of Physical Anthropology* **88**: 469–82.

Garber, P.A. (1993). Feeding ecology and behaviour of the genus *Saguinus*. In: *Marmosets and Tamarins: Systematics, Behaviour, and Ecology*, A.B. Rylands, ed., Oxford: Oxford University Press. pp. 273–95.

Goldizen, A.W. (1987). Tamarins and marmosets: Communal care of offspring. In: *Primate Societies*, B.B. Smuts, D.L. Cheney, R.M. Seyfarth, R.W. Wrangham and T.T. Struhsaker, eds., Chicago: University of Chicago Press, pp. 34–43.

Hardie, S.M. (1998). Mixed-species tamarin groups (*Saguinus fuscicollis* and *Saguinus labiatus*) in northern Bolivia. *Primate Report* **50**: 39–62.

Hardie, S.M., Day, R.T., and Buchanan-Smith, H.M. (1993). Mixed-species *Saguinus* groups at Belfast Zoological Gardens. *Neotropical Primates* **1**: 19–21.

Hauser, M.D., and Wrangham, R.W. (1990). Recognition of predator and competitor calls in nonhuman primates and birds: A preliminary report. *Ethology* **86**: 116–30.

Hershkovitz, P. (1977). *Living New World Monkeys (Platyrrhini)*, Vol 1. Chicago: University of Chicago Press.

Heymann, E.W. (1987). A field observation of predation on a moustached tamarin (*S. mystax*) by an anaconda. *International Journal of Primatology* **8**: 193–5.

Heymann, E.W. (1990). Reactions of wild tamarins *Saguinus mystax* and *Saguinus fuscicollis* to avian predators. *International Journal of Primatology* **11**: 327–37.

Heymann, E.W., and Buchanan-Smith, H.M. (2000). The behavioural ecology of mixed-species troops of callitrichine primates. *Biological Reviews of the Cambridge Philosophical Society* **75**: 169–90.

Izawa, K. (1978). A field study of the ecology and behaviour of the black-mantled tamarin *Saguinus nigricollis*. *Primates* **19**: 241–74.

Leopold, A.S. (1959). *Wildlife of Mexico*. Berkeley: University of California Press.

MacDonald, S.E., Pang, J.C., and Gibeault, S. (1994). Marmoset (*Callithrix jacchus jacchus*) spatial memory in a foraging task – Win–stay versus win–shift strategies. *Journal of Comparative Psychology* **108**: 328–34.

Macedonia, J.M., and Yount, P.L. (1991). Auditory assessment of avian predator threat in semicaptive ringtailed lemurs (*Lemur catta*). *Primates* **32**: 169–82.

Martin, P., and Bateson, P. (1986). *Measuring Behaviour*. Cambridge: Cambridge University Press.

Moynihan, M. (1970). Some behavior patterns of platyrrhine monkeys: II. *Saguinus geoffroyi* and some other tamarins. *Smithsonian Contributions to Zoology* **28**: 1–77.

Napier, J.R., and Napier, P.H. (1967). *A Handbook of Living Primates*. New York: Academic Press.

Neyman, P.F. (1978). Aspects of the ecology and social organisation of free-ranging cotton-top tamarins (*Saguinus oedipus*) and the conservation status of the species. In: *Biology and Conservation of the Callitrichidae*, D.G. Kleiman, ed., Washington DC: Smithsonian Institution Press, pp. 39–71.

Norconk, M.A. (1990). Mechanisms promoting stability in mixed *Saguinus mystax* and *S. fuscicollis* troops. *American Journal of Primatology* **21**: 129–46.

Peres, C.A. (1991). Ecology of mixed species groups of tamarins in Amazonian *terra firme* forests. Unpublished Ph.D. thesis, University of Cambridge, Cambridge.

Peres, C.A. (1993). Anti-predator benefits in a mixed-species group of Amazonian tamarins. *Folia Primatologica* **61**: 61–76.

Pook, A.G., and Pook, G. (1982). Polyspecific association between *Saguinus fuscicollis*, *Saguinus labiatus*, *Callimico goeldii*, and other primates in north-western Bolivia. *Folia Primatologica* **38**: 196–216.

Prescott, M.J. (1999). Social learning in mixed-species troops of *Saguinus fuscicollis* and *Saguinus labiatus*: Tests of foraging benefit hypotheses in captivity. Unpublished Ph.D. thesis, University of Stirling, Stirling.

Prescott, M.J., and Buchanan-Smith, H.M. (1999). Intra- and inter-specific social learning of a novel food task in two species of tamarin. *International Journal of Comparative Psychology* **12**: 1–22.

Ramirez, M. (1989). Feeding ecology and demography of the moustached tamarin, *Saguinus mystax*, in northeastern Peru. Unpublished Ph.D. thesis, City University of New York, New York.

Snowdon, C.T., and Soini, P. (1988). The tamarins, genus *Saguinus*. In: *Ecology and Behavior of Neotropical Primates*, Vol 2, R.A. Mittermeier, A.B. Rylands, A. Coimbra-Filho and G.A.B. de Fonesca, eds., Washington DC: World Wildlife Fund, pp. 223–98.

Soini, P. (1987). Ecology of the saddle-back tamarin, *Saguinus fuscicollis illigeri* on the Rio Pacaya, northeastern Peru. *Folia Primatologica* **49**: 11–32.

Stafford, B.J., and Ferreira, F.M. (1995). Predation attempts on callitrichids in the Atlantic coastal rain forest of Brazil. *Folia Primatologica* **65**: 229–33.

Sussman, R.W., and Kinzey, W.G. (1984). The ecological role of the Callitrichidae. *American Journal of Physical Anthropology* **64**: 419–49.

Terborgh, J. (1983). *Five New World Primates: A Study in Comparative Ecology*. Princeton: Princeton University Press.

Terborgh, J. (1990). Mixed flocks and polyspecific associations: Cost and benefits of mixed groups to birds and monkeys. *American Journal of Primatology* **21**: 87–100.

Yoneda, M. (1981). Ecological studies of *Saguinus fuscicollis* and *S. labiatus* with reference to habitat segregation and height preference. Kyoto University Overseas Research Report of New World Monkeys, II, 43–50.

Yoneda, M. (1984). Comparative studies on vertical separation, foraging behavior and travelling mode of saddle-backed tamarins (*Saguinus fuscicollis*) and red-chested moustached tamarins (*Saguinus labiatus*) in Northern Bolivia. *Primates* **25**: 414–22.

Zuberbühler, K., Noë, R., and Seyfarth, R.M. (1997). Diana monkey long-distance calls: Messages for conspecifics and predators. *Animal Behaviour* **53**: 589–604.

4 • Seeing red: Consequences of individual differences in color vision in callitrichid primates

NANCY G. CAINE

Introduction

Behavioral adaptations are mediated by the sensory systems. Animals locate food, identify mates, and avoid predators when they see, hear, smell, or otherwise sense them. It is easy to take for granted the fact that the sensory systems are directly and strongly influenced by selection pressures, but the identification of those pressures and the resulting adaptations sheds light on behavior we wish to understand. In this chapter the ways in which individual differences in one particular sensory adaptation, color vision, may influence foraging and predator detection in callitrichid primates are described. It is argued here that predation sensitive foraging may be accomplished in ways unique to callitrichids and other primate species that display this interesting sensory polymorphism.

Sensory specialization and compromise
Sensory systems often reflect a compromise resulting from different, sometimes opposing, selection pressures. A familiar and excellent example of this fact is the response of the visual system to selection pressures associated with diurnal versus nocturnal life. When light is plentiful, the eye can afford to specialize in the detection of detail in the visual world. In the absence of light, the eye must do what it can to capture and respond to every bit of illumination, even though such sensitivity sacrifices the visual detail enjoyed by diurnal species. Diurnal and nocturnal eyes are different in a host of ways, but most fundamentally they differ in the relative number of the two types of photoreceptors: rods and cones. Rods have a lower threshold to light than do cones and thus function when cones do not. Cones are of little use under conditions of low illumination but they do what rods cannot: respond

differentially to wavelength. To the extent that a species specializes in diurnal or nocturnal behavior, the visual system is correspondingly committed to a high or low cone/rod ratio, along with other related adaptations; highly specialized diurnal species do not function well in the dark (at least in terms of vision) and nocturnal species show relatively poor visual acuity regardless of light level (Levine 2000). Behavior is constrained in accordance with these basic characteristics of the visual system.

Most primates not only have retinas dense with cones, but they also have different kinds of cones, the result of which is trichromatic (three different cone opsins) color vision (explained in more detail below). It is generally believed that primates enjoy a foraging advantage related to their trichromacy (e.g., Mollon 1989). Specifically, the particular light absorption curves of primate cones seem rather ideally designed for the task of detecting ripe fruit against dappled green foliage (Jacobs 1995, Nagle and Osorio 1993). This claim has been supported with a number of studies that measure the spectroradiometric properties of natural backgrounds and fruits (Osorio and Vorobyev 1996, Regan *et al.* 1996). These studies confirm the likely superiority of trichromats under natural fruit foraging conditions. Recently, Lucas *et al.* (1998) and Dominy and Lucas (2001) have provided evidence that the value of trichromacy for certain leaf-eating species relates more to enhanced detection of palatable leaves, which are often dappled red.

Although the literature on the adaptive significance of primate color vision focuses almost exclusively on *foraging* advantages associated with trichromacy, an obvious extension of the argument is that predator detection is similarly affected by the ability of an individual to make use of its visual perception abilities. If a predator's coloration provides a contrast against the background where that predator happens to be, and if that color contrast involves the relevant wavelengths (e.g., an orange snake against a green background), the trichromat might have an advantage in detecting the threat.

As stated above with regard to diurnal and nocturnal eyes, sensory specialization often entails a cost or compromise. Trichromatic color vision confers certain visual advantages, but are trichromats visually inferior to dichromats (two cone opsins) or monochromats (one cone opsin) in any important ways? In 1943, Deane B. Judd of the National Bureau of Standards published an article in *Science* that was, apparently, a response to newspaper reports claiming that 'colorblind observers have frequently been successful in spotting otherwise perfectly camouflaged positions' (Judd 1943, p. 544). Judd concludes that, in fact, certain camouflaged objects would be more difficult for trichromats than dichromats (color 'blind' people) to detect. Morgan *et al.* (1992) describe the perceptual basis for this disadvantage as follows:

> It is an interesting feature of our visual system that we can entertain only one perceptual organization at once, rather as we cannot simultaneously instruct our limbs to flex and extend . . . we propose that when alternative methods of segmenting come into competition, one of the potential organizations will be selected at the expense of the others. (p. 294)

If humans tend to give perceptual priority to color, then a trichromat, when faced with the task of detecting a shape when that shape is confounded by color, will respond first to the color cues, thereby handicapping him or herself in the task of detecting the object. This prediction was tested by Morgan *et al.* (1992). The stimulus in their study was an array of small rectangles (or, in a second condition, capital As and Bs), some of which were red and others of which were green. The rectangles were arranged in such a way that a patch of them differed in orientation from the others, creating a target area of different 'texture.' The subjects' task was to identify, within 200 ms, the location of that target area within the array. If, as predicted, trichromats give priority to the analysis of color, then their success at detecting the figure would be less than those of dichromats, for whom color was not a confounding feature of the array. The results were unambiguous: the dichromats correctly located more of the figures than did the trichromats. Furthermore, there was no difference between the groups when the rectangles were red and blue, or green and blue: dichromats do see blue, and hence they, like the trichromats, were handicapped by the confounding feature of color. There was no difference in how often the trichromats and dichromats located figures that were uniformly colored.

These results are the first (and, to date, the only) empirical, published demonstration of the superior ability of dichromats to penetrate color camouflage. If this advantage generalizes from the laboratory to the field, and from humans to nonhumans, a case can be made that dichromats may detect some kinds of food and predators better than trichromats. This, in turn, might explain why the dichromat phenotype persists at low but 'non-negligible' rates among humans (Verhulst and Maes 1998, p. 3387). However, in most species (including humans) it appears that almost all individuals are either dichromat (most mammals) or trichromat (most primates), offering no basis for comparison of foraging or predator detection abilities across individuals in naturalistic settings. The unusual case of New World monkeys provides us an opportunity to make precisely this comparison. Below, a brief review of the physiology of color vision in primates is presented as background for appreciating the research opportunities posed by the visual polymorphism of many New World species.

Primate visual systems: color vision

As a taxon, primates are known for their excellent vision, although, of course, monkeys and apes also have acute auditory abilities and many species are also highly sensitive to olfactory stimuli. Most simian primates are strongly diurnal and their sensory physiology reflects that fact. Diurnal primate retinas are cone-rich and primate visual neurophysiology is specialized for the interpretation of visual detail (color, shape, texture, etc.) (Tovee 1996).

Cones, like rods, contain visual pigments (opsins), chemicals that absorb light. Each cone contains an opsin that is maximally responsive to a particular wavelength of light, the result being a distinct pattern of neural activity in the retina depending upon what wavelength(s) of light are available and which opsins are present. The pattern of neural activity associated with three different cone opsins is the basis for our ability to see color as we know it. The vast majority of mammals have two opsins, one that responds best to relatively short wavelengths (light we would call blue) and one that responds best to a wavelength in the mid-to-long part of the spectrum (green, yellow, red) (Jacobs 1993). We call such species 'dichromatic,' and their perception of color is limited. Humans who we call 'color blind' are typically dichromats; they see blue and yellow, but reds and greens are indistinguishable from each other and appear gray/brown/yellow, depending on the exact wavelength and viewing conditions. Most mammals, then, are like color 'blind' humans; a few mammalian species are more truly color blind, lacking cones or having only one type of opsin in those cones (Jacobs 1993).

Old World monkeys, apes, and humans are trichromats: we have three types of cones (opsins) in our retinas. The three opsins include one that is maximally sensitive to about 430 nm (and hence called the short-wavelength, SW, opsin) and two that are in the mid–long (M/L) wavelength category (about 530 nm and 560 nm). It is now known that many humans actually have more than three types of cones, but our perceptions of color are probably very similar to those of Old World primates (Jacobs 1993, Neitz and Jacobs 1990, Neitz *et al.* 1993). New World monkeys and some prosimians (Tan and Li 1999) present a most interesting and unusual case of color vision, the behavioral consequences of which are the focus of this chapter.

The production of opsins is dependent upon the presence and expression of genes that code specifically for those opsins (Nathans *et al.* 1986). In primates, the gene for the SW opsin is autosomal and monomorphic. Old World monkeys, apes and humans have two additional genes on the X chromosome that code for the MW and LW opsins. Thus, all normal individuals in these primate taxa have three different opsins, allowing them to see the range of colors we know as the visible spectrum.

The unusual case of New World monkey color vision

In 1984, Jacobs published research on squirrel monkey color vision that suggested a rather startling pattern of individual differences in that species. Some individuals were clearly dichromats; they were able to make limited behavioral color discriminations consistent with the presence of two opsins. Other squirrel monkeys, however, made discriminations that require the presence of three opsins. The trichromats were always female, whereas the dichromats included both females and males. Furthermore, it was determined that there were actually six distinct color vision phenotypes: three trichromat types and three dichromat types (Bowmaker *et al.* 1987, Jacobs 1984). Developments in opsin genetics and the molecular structure of photopigments revealed the basis of this polymorphism. As is the case for other primates, squirrel monkeys have an autosomal gene that codes for the SW pigment. Unlike other simian primates, however, these platyrrhines have one, not two, loci on the X chromosome that codes for a M/L wavelength opsin. The M/L gene comes in three forms, ranging from about 535 nm to 562 nm in peak sensitivity (Jacobs *et al.* 1993). If a female inherits a different form of the M/L gene on each of her two X chromosomes, then she will be a trichromat. The same allele on both of her X chromosomes would make her a dichromat. Males, of course, can only be dichromats. We now know that this arrangement is also present in callitrichids (with slightly different peak absorption values) and most other cebids. Howler (genus *Alouatta*) monkeys, being habitually trichromatic, are the intriguing exception (Jacobs *et al.* 1996). Their trichromacy may have to do with their dietary dependence on leaves at certain times of the year (Lucas *et al.* 1998).

It is important to remember that dichromats (most mammals) do in fact have color vision; recall that the foundation of color vision is the differential response of opsins to wavelength and having two opsin types means that the nervous system can, given the right 'wiring,' make a comparison of those responses. It is estimated that, about 30 million years ago, a second opsin gene locus was added to the X chromosome in an early Old World primate species (Nathans *et al.* 1986, Yokoyama and Yokoyama 1990). Jacobs (1995) explains that a crossover event involving two different alleles could have placed two opsin genes on one chromosome, or perhaps the opsin gene on the X chromosome was duplicated and subsequently mutated. Either way, by differing from each other by just a few amino acids, the two genes produce opsins whose peak absorption profiles are different than each other. Presumably, the addition of a third spectral absorption curve for comparison with the other two conferred a powerful adaptive advantage on its bearer: the ability to distinguish between green and red.

The ability to make distinctions among wavelengths in the mid-to-long ranges generates a powerful prediction regarding the likely superiority of trichromats in certain foraging situations. However, it also invites the examination of a question that has been posed before: if trichromacy is so adaptive, why does the dichromat phenotype remain in the population? One explanation has to do with the fact that routine trichromacy depends upon the establishment of a second opsin locus on the X chromosome (see above). Perhaps this genetic event is so unusual as to have simply not yet happened in callitrichids, although the existing polymorphism seems to be quite ancient (Boissinot *et al.* 1998, Shyue *et al.* 1995). Jacobs *et al.* (1996) make the case that howler monkeys, the only New World genus currently known to be routinely trichromatic, represent such an unusual event, having established the additional locus sometime after the development of the polymorphic condition of the other platyrrhines. In comparison, based on laboratory data from humans, we have some reason to predict that dichromats may fare better than trichromats under conditions of low light (Verhulst and Maes 1998) and color camouflage (Morgan *et al.* 1992). If so, there could be active selection for the dichromat phenotype.

Until recently, however, none of the existing data on the adaptive significance of trichromatic vision directly test the hypothesis of trichromat advantage, namely, that trichromats forage more successfully than dichromats in situations involving ripe fruit against a green background. Likewise, there are no direct tests of the hypothesis that dichromats might excel in detecting food – or predators – where color camouflage is involved. The following two studies, one recently published (Caine and Mundy 2000) and one recently completed, take advantage of the platyrrhine polymorphism in color vision to test the hypothesis that trichromat marmosets have an advantage in detecting orange items against a background of green, whereas dichromats are better able to detect items that are color camouflaged. Although these two studies use food items as stimuli, the hypotheses could be applied equally well to nonfood items in the animals' environments. It will be argued that trichromats and dichromats have different and complementary visual specializations that are applied to the detection of many salient features of their environments. To the extent that the relevant color patterns apply to both predators and prey, predation sensitive foraging in callitrichids can be investigated from the perspective of individual differences in visual abilities.

Method: Study 1

At the Center for Reproduction of Endangered Species (CRES), San Diego Wild Animal Park, two groups of Geoffroy's marmosets (*Callithrix geoffroyi*) live in large outdoor enclosures. The floors of the enclosures are covered in naturally-occurring weeds, grasses, and dirt, and the monkeys spend up to 60% of their time foraging for insects and pieces of provisioned fruits that have fallen to the ground (Caine 1996). This situation presents an opportunity to observe foraging success in a naturalized environment while maintaining some amount of experimental control.

To create a situation in which the marmosets could forage for food that would test the theory of trichromat advantage, individual pieces of Kix® cereal, which are tan-colored, round, and approximately 1 cm. in diameter, were sprayed with food color to create either an orange or green appearance. There was no attempt to standardize precisely the process of coloring the Kix®; rather, our goal was to mimic the natural situation wherein fruit or other food items vary to some degree. Therefore, some pieces were darker, or more uniformly shaded, than others, but to a human trichromat, all orange Kix® were distinctly different than all green Kix®. A color blind human male was also asked to look at the dyed Kix®. He reported that the orange and green Kix® were generally indistinguishable from each other and from the backgrounds of dirt and grass in the enclosures.

On each of 45 days, orange Kix® were broadcast across the marmoset enclosure floors. On each of 45 alternate days, green Kix® were broadcast. Kix® are highly palatable to the marmosets and thus the monkeys were very motivated to search for them. In each trial the marmosets were observed for 10 minutes. Every time a Kix® was found, the identity of the marmoset that found it and the elapsed time since the beginning of the trial were recorded. In this experiment, the marmosets searched for the Kix® from distances up to six meters (the width of the enclosures).

While the behavioral data were being collected, the 14 marmosets in the experiment were genotyped. DNA was extracted from plucked hairs using either a Chelex method (Garza and Woodruff 1992) or a commercial DNA extraction kit (Qiagen tissue extraction kit). (For details of this procedure, see Caine and Mundy 2000.) All sites thought to be important for determining spectral sensitivity were sequenced, and it was determined that (a) there were both dichromat and trichromat females in the CRES population; (b) the genotypes were consistent with the known pedigree of each animal; (c) the three M/L opsins are identical to those determined for common marmosets, *Callithrix jacchus* (543 nm, 556 nm, 563 nm) (Hunt *et al.* 1993, Travis *et al.* 1988). Thus, in addition to allowing for an analysis of the behavioral data according to genotype, we were able to add another species to the list of New World monkeys that exhibit the polymorphism described above.

Because I collected the behavioral data and my coauthor, Nick Mundy, did the DNA analyses, I was able to remain blind to the genotype of the monkeys as I ran the trials. Once behavioral data collection was completed, we analyzed the data according to the following hypotheses: (a) trichromats would find more orange than green Kix®; (b) dichromats would not differ in the number of Kix® they found of each color; (c) trichromats would find more orange Kix® than dichromats but they would find equal numbers of green Kix®. The basis for these predictions was that the orange Kix® would, to a trichromat, stand out against the predominantly green foliage of the enclosures. There would be no such contrast for the dichromats. For both dichromats and trichromats the green Kix® would blend in with the foliage, presenting, for the trichromats, a

Table 4.1. *Numbers of Kix® found by dichromat and trichromat marmosets in Study 1*

	Number of orange Kix® found	Number of green Kix® found
Trichromats		
Ca	27	16
Sh	77	50
Go	37	31
To	18	14
Ys	31	14
Db	32	22
Dichromats		
Pa (male)	16	20
Gr	24	36
Mo (male)	26	27
Mk	12	12
Yz (male)	28	30
Sch (male)	27	28
So	11	19
Dv (male)	15	13

Source: After Caine and Mundy 2000.

more difficult task than finding orange Kix®. For the dichromats, who do not distinguish orange from green, the two conditions present the same challenge. Wilcoxon tests and Mann–Whitney tests were used for within group and between group comparisons, respectively ($p < 0.05$).

Results: Study 1

As shown in Table 4.1, the results were statistically significant in accordance with all three predictions. On average, trichromats (of which there were six in our sample) found 20.3% more orange than green Kix® and all of them found at least 9% more orange than green. Most of the dichromats found very similar numbers of orange and green Kix®. Of the three females in our sample who were dichromats, none found more orange than green, eliminating the possibility that it was the property of being female, rather than being a trichromat, that accounts for the greater success in foraging for orange Kix®. Trichromats and dichromats found similar numbers of green Kix® but trichromats found significantly more orange Kix® than did the dichromats.

Method: Study 2

Having shown that trichromat marmosets were indeed superior to dichromat marmosets at finding orange food against a dappled green background, we next wished to assess the prediction that dichromat marmosets would forage more successfully than trichromats under conditions of color camouflage. On each of 45 days over the course of 4 months, orange (22 trials) or green (23 trials) Kix® were hidden in clear plastic tubs (18 cm diameter × 10.5 cm. deep) filled with like-colored shavings. The tubs were then placed in various areas around the floor of the enclosures where the marmosets typically forage. On an additional 45 days, alternating with orange or green trial days, Kix® painted both orange and green were hidden in tubs filled with a mixture of orange and green shavings. Three and six tubs were used for groups 1 and 2, respectively, and three Kix® were hidden in each tub. The marmosets were free to sift though the shavings in any of the various tubs for up to 10 minutes, and the number of Kix® found by each marmoset was recorded.

Results: Study 2

As with Study 1, Wilcoxon and Mann–Whitney tests were used for data analysis ($p<0.05$). In accordance with our prediction, and as shown in Table 4.2, trichromats found significantly more solid-color than mixed-color Kix® but dichromats found solid- and mixed-color Kix® equally often. On average, trichromats found 24.9% fewer mixed-color than solid-color Kix® and all of the trichromats found at least 11% fewer mixed than solid. Between groups tests revealed that trichromats (mean, $\bar{X}=35.6$; SE$=8.2$) and dichromats ($\bar{X}=39.9$, SE$=9.2$) found similar numbers of solid Kix®. On average, trichromats found many fewer mixed-color Kix® than did the dichromats ($\bar{X}=21.4$, SE$=5.7$; $\bar{X}=39.2$, SE$=10.1$, respectively), but the difference was not statistically significant. Given the wide range of individual differences in total number of Kix® found both within and between trichromats and dichromats, the failure to reach statistical significance in the between subjects comparisons is not surprising.

Table 4.2. *Numbers of Kix® found by dichromat and trichromat marmosets in Study 2*

	Number of mixed-color Kix® found	Number of solid-color Kix® found
Trichromats		
Ca	22	43
Go	42	58
To	14	25
Ys	08	10
Db	21	42
Dichromats		
Pa (male)	12	10
Gr	97	87
Mo (male)	15	21
Mk	11	13
Yz (male)	46	41
Sch (male)	90	82
So	08	8
Dv (male)	43	51
Em (male)	26	26
Es	44	60

Discussion

Our data indicate that, under certain lighting and background conditions, trichromats may excel at locating objects that are in the red/orange part of the spectrum but may be handicapped in locating items that are red/green color camouflaged. Although the comparison of dichromats and trichromats in Study 2 (camouflage) did not reach statistical significance, the trend was clearly in the right direction, and the trichromats did display a deficit with regard to their own success at finding camouflaged versus noncamouflaged items. Trichromat and dichromat marmosets may, therefore, occupy different visual niches.

Although the results presented above were based on studies using food, there is no reason to think that the differential advantages noted for trichromats and dichromats would not apply to predators whose coloration makes them stand out against dappled foliage or blend in against a mottled background of reds and greens. What is perhaps most intriguing about the consequences of color-vision polymorphism in callitrichids is the possibility that individuals within a group can act as 'specialists' in detecting certain kinds of food or predators, while other members of their groups can specialize in detecting other varieties of food or predators. Callitrichids live in fairly small (4–12 individuals, typically) groups that are remarkably cooperative, cohesive, and vigilant. All members of a group carry and protect infants, sleep together in a communal huddle, participate in group-mediated vigilance, and often cooperate at food sources (Schaffner and Caine 2000). It is generally agreed that the within-group cooperation that so characterizes callitrichids is a necessary set of adaptations related to short interbirth intervals, twinning, and heavy predation pressure (Caine 1993, Garber 1997). Callitrichids must travel considerable distances to locate their primary food sources: fruit, insects, and (especially for marmosets) exudates from trees. The process of locating and consuming their food must, of course, be balanced with the need to maintain vigilance for aerial and terrestrial predators, which come in many forms. From the perspective of sensory abilities, no one set of specializations can optimally respond to all possible foraging and antipredatory challenges. Dichromat marmosets may enjoy certain advantages over trichromats, such as the ability to detect color camouflaged food or predators (Morgan *et al.* 1992), or, as has recently been shown in humans, see better at dawn and dusk (Verhulst and Maes 1998); trichromats may more easily find ripe fruit (Mollon 1989, Caine and Mundy 2000). In this scenario, neither phenotype enjoys a large advantage over the other, especially when the two live cooperatively in groups.

The notion that individual differences in the visual abilities of callitrichids help to explain tamarin and marmoset social cooperation and social roles generates many testable hypotheses. For instance, callitrichids use sentinels in some circumstances. One situation in which sentinels are sometimes apparent (in both captivity and the wild) is prior to entering the sleeping site in the evening (Zullo and Caine 1988). Given the recent evidence that dichromat humans have lower light detection thresholds than trichromats (Verhulst and Maes 1998), we might predict that dichromat callitrichids more commonly assume the role of pre-retirement sentinel than do their trichromat groupmates. Similarly, we might predict (as suggested by Mollon 1989) that trichromat females are more likely to lead the way to distant fruit sources, which they might see before their dichromat groupmates do. Group differences may also be studied in terms of the color-vision phenotypes represented. Might callitrichid groups that contain only dichromats fare more poorly than groups with both phenotypes in terms of foraging, predator avoidance, and reproductive success?

Recognition of the fact that these are plausible hypotheses also requires us to reconsider some explanations we have derived about callitrichid individual or group behavior. For instance, males are sometimes reported to be more vigilant than females (e.g., Koenig 1998). However, it is possible that the correct explanation is that *dichromats* are more vigilant than *trichromats* with sex being only a correlated variable. Other individual differences in behavior related to predation and foraging, heretofore unexplained, may also become understood if the visual capabilities of the individuals are closely examined.

Summary

As is amply demonstrated in this volume, primates that are subject to predation balance the demands of foraging with the need to avoid contact with predators. Although we tend to explain predation sensitive foraging at the level of the behavior itself (e.g., the formation of mixed-species groups or the assignment of vigilance roles to certain group members), the ability of animals to forage and avoid predators successfully is ultimately dependent on the sensory mechanisms used to detect relevant environmental stimuli. In the visual system, the ability to discriminate among wavelengths of light (i.e., to see colors) confers some potent advantages – but also, perhaps, some disadvantages – on primates. Color can alert a forager to the distant location of a ripe piece of fruit in a tree, but a brain that gives priority to the analysis of color cues can obscure from the viewer some textural and shape cues. As is often the case within the sensory systems, specialization of function involves a trade-off; to the extent that animals can associate with allies that have complementary abilities, the costs of specialization are reduced.

In contrast to the vast majority of mammalian taxa, which are dichromatic, most primate taxa are trichromatic. However, callitrichids, most cebids, and at least some prosimians have both dichromats and trichromats in their populations. In these taxa, owing to the location and number of genes coding for photopigments, all males are dichromats, but some females are trichromats. The accumulation of data supporting the contention that trichromacy is associated with a foraging advantage (the ability to see ripe fruit or nutritious leaves) has generated curiosity about the retention of the dichromat phenotype in New World monkeys. One possibility is that there are one or more distinct advantages that dichromats have over trichromats; based on data from humans, these advantages may include a lower threshold of light detection and the ability to penetrate color camouflage. There is now evidence that Geoffroy's marmosets, like other callitrichids studied to date, include both trichromat and dichromat phenotypes; in accordance with prediction, the trichromats are better able to find orange food against a green background, but may be disadvantaged in finding color-camouflaged food. It seems reasonable to assume that these complementary abilities would extend to the realm of predator detection, as well. When wild marmoset groups contain members of both phenotypes, the task of finding food while avoiding predators may well be facilitated by individual differences in sensory abilities.

REFERENCES

Boissinot, S., Tan, Y., Shyue, S.-K., Schneider, H., Sampaio, I., Neiswanger, K., Hewett-Emmett, D., and Li, W.-H. (1998). Origins and antiquity of X-linked triallelic color vision systems in New World monkeys. *Proceedings of the National Academy of Sciences* **95**: 13 749–54.

Bowmaker, J.K., Jacobs, G.H., and Mollon, J.D. (1987). Polymorphism of photopigments in the squirrel monkey: a sixth phenotype. *Proceedings of the Royal Society of London, Series B*, **231**: 383–90.

Caine, N.G. (1993). Flexibility and cooperation as unifying themes in *Saguinus* social organization and behaviour: the role of predation pressures. In: *Marmosets and Tamarins: Systematics, Behaviour, and Ecology*, A.B. Rylands, ed., Oxford: Oxford University Press, pp. 200–19.

Caine, N.G. (1996). Foraging for animal prey by outdoor groups of Geoffroy's marmosets (*Callithrix geoffroyi*). *International Journal of Primatology* **17**: 933–45.

Caine, N.G., and Mundy, N. I. (2000). Demonstration of a foraging advantage for trichromatic marmosets (*Callithrix geoffroyi*) dependent on food colour. *Proceedings of the Royal Society of London, Series B* **267**: 439–44.

Dominy, N.J., and Lucas, P.W. (2001). Ecological importance of trichromatic vision to primates. *Nature* **410**: 363–6.

Garber, P.A. (1997). One for all and breeding for one: cooperation and competition as a tamarin reproductive strategy. *Evolutionary Anthropology* **5**(6): 187–99.

Garza, J.C., and Woodruff, D.S. (1992). A phylogenetic study of the gibbon (*Hylobates*) using DNA obtained non-invasively from hair. *Molecular and Phylogenetic Evolution* **1**: 202–10.

Hunt, D.M., Williams, A.J., Bowmaker, J.K., and Mollon, J.D. (1993). Structure and evolution of the polymorphic photopigment gene of the marmoset. *Vision Research* **33**(2): 147–54.

Jacobs, G.H. (1984). Within-species variations in visual capacity among squirrel monkeys (*Saimiri sciureus*): color vision. *Vision Research* **24**: 1267–77.

Jacobs, G.H. (1993). The distribution and nature of colour vision among the mammals. *Biological Review* **68**: 413–71.

Jacobs, G.H. (1995). Variations in primate color vision: mechanisms and utility. *Evolutionary Anthropology* **3**(6): 196–205.

Jacobs, G.H., Neitz, J., and Neitz, M. (1993). Genetic basis of polymorphism in the color vision of platyrrhine monkeys. *Vision Research* **33**: 269–74.

Jacobs, G.H., Neitz, M., Deegan, J.F., and Neitz, J. (1996). Trichromatic colour vision in New World monkeys. *Nature* **382**: 156–8.

Judd, D.B. (1943). Colorblindness and the detection of camouflage. *Science* **97**: 544–6.

Koenig, A. (1998). Visual scanning by common marmosets (*Callithrix jacchus*); functional aspects and the special role of adult males. *Primates* **39**(1): 85–90.

Levine, M.W. (2000). *Fundamentals of Sensation and Perception*. Oxford: Oxford University Press.

Lucas, P.W., Darvell, B.W., Lee, P.K.D., Yuen, T.D.B., and Choong, M.F. (1998). Colour cues for leaf food selection by long-tailed macaques (*Macaca fascicularis*) with a new suggestion for the evolution of trichromatic colour vision. *Folia Primatologica* **69**: 139–52.

Mollon, J.D. (1989). 'Tho' she kneel'd in that place where they grew . . .' The uses and origins of primate colour vision. *Journal of Experimental Biology* **146**: 21–38.

Morgan, M.J., Adam, A., and Mollon, J.D. (1992). Dichromats detect colour-camouflaged objects that are not detected by trichromats. *Proceedings of the Royal Society of London, Series B* **248**: 291–5.

Nagle, M.G., and Osorio, D. (1993). The tuning of human photopigments may minimize red-green chromatic signals in natural conditions. *Proceedings of the Royal Society of London, Series B* **252**: 209–13.

Nathans, J., Thomas, D., and Hogness, D.S. (1986). Molecular genetics of human color vision: the genes encoding blue, green, and red pigments. *Science* **232**: 193–202.

Neitz, J., and Jacobs, G.H. (1990). Polymorphism in normal human color vision and its mechanism. *Vision Research* **30**(4): 621–36.

Neitz, J., Neitz, M., and Jacobs, G.H. (1993). More than three different cone pigments among people with normal color vision. *Vision Research* **33**: 117–22.

Osorio, D., and Vorobyev, M. (1996). Colour vision as an adaptation to frugivory in primates. *Proceedings of the Royal Society of London, Series B* **263**: 593–9.

Regan, B.C., Vienot, F., Charles-Dominique, P.C., Pefferkorn, S., Simmen, B.,

Julliot, C., and Mollon, J.D. (1996). The colour signals that fruits present to primates. *Investigative Ophthalmology and Visual Sciences* **37**(3): S648.

Schaffner, C.M., and Caine, N.G. (2000). The peacefulness of cooperatively breeding primates. In: *Natural Conflict Resolution*, F. Aureli and F.B.M De Waal, eds., Berkeley: University of California Press, pp. 155–69.

Shyue, S-K., Hewett-Emmett, D., Sperling, H.G., Hunt, D.G., Bowmaker, J.K., Mollon, J.D., and Li, W.-H. (1995). Adaptive evolution of color vision genes in higher primates. *Science* **269**: 1265–7.

Tan, Y., and Li, W.-H. (1999). Trichromatic vision in prosimians. *Nature* **402**: 36.

Tovee, M.J. (1996). *An Introduction to the Visual System*. Cambridge: Cambridge University Press.

Travis, D.S., Bowmaker, J.K., and Mollon, J.D. (1988). Polymorphism of visual pigments in a callitrichid monkey. *Vision Research* **28**: 481–90.

Verhulst, S., and Maes, F.W. (1998). Scotopic vision in colour-blinds. *Vision Research* **38**: 3387–90.

Yokoyama, R., and Yokoyama, S. (1990). Convergent evolution of the red- and green-like visual pigment genes in fish, *Astyanax fasciatus*, and human. *Proceedings of the National Academy of Science* **87**: 9315–18.

Zullo, J., and Caine, N.G. (1988). The use of sentinels in captive groups of red-bellied tamarins. *American Journal of Primatology* **14**: 455.

5 • Predator sensitive foraging in Thomas langurs

ELISABETH H.M. STERCK

Introduction

The behavior of primates results from a compromise of requirements and risks. The requirements include the acquisition of food and mates and the risks include predators and dangerous conspecifics. Both the acquisition of food and the risk of predation will affect behavior on a daily basis. Food, however, may be found on potentially dangerous locations and foraging behavior may be influenced by strategies to reduce predation risk. This paper examines whether this is found for Thomas langurs (*Presbytis thomasi*).

Many diurnal primates live in groups to reduce the risk of predation (Alexander 1974; Janson and Goldsmith 1995; van Schaik 1983). Individuals in these groups may benefit from group members in several ways: the early detection of predators, alarm calls, confusion of the predator, the selfish herd effect and by joint mobbing and dislodging of the predator (Alcock 1998). A larger group size or more neighbors (Treves 1998) may provide better protection. Not all group members, however, may benefit equally and high-ranking individuals may occupy safer places within the group than low-ranking ones (e.g., Janson 1990a,b; van Schaik and van Noordwijk 1988).

Langur species are largely arboreal primates that are confined to Asia. Their predators include snakes, felids and canids. Snakes and the smaller felids can be found at all heights in the forest, but are probably less agile in trees than primates (e.g., leopards (*Panthera leo*): Muckenhirn 1972; personal observation of encounters of primates with reticulated pythons (*Python reticularis*) and a clouded leopard (*Neofelis nebulosa*)), whereas the tiger (*Panthera tigris*) and canids (jackals (*Canis aureus*): Stanford 1991; domestic dogs (*Canis domesticus*): Sommer 1985) are confined to the ground. These are all predators that have to come relatively close (i.e., the attack distance of leopards on the ground is probably 8–20 m, Muckenhirn 1972; jackals: 10 m, Stanford 1991; for predators in trees it is probably 2–10 m, based on observed mobbing distances) to their prey before they can an attack (Stanford 1991; Steenbeek *et al.* 1999; Sterck 1996, 1997). No raptors that are capable of taking adult langurs are sympatric with langurs in Asia (e.g., van Schaik and Hörstermann 1994). Therefore, the treetops and canopy are not particularly dangerous. Altogether, the ground and possibly also the understory are the locations where the predation risk will be largest and predation risk will mainly affect behavior on or close to the ground (Hanuman langurs (*Presbytis entellus*): Yoshiba 1968 in Muckenhirn 1972; Thomas langurs: Steenbeek *et al.* 1999; Sterck 1995).

The type of predator that Thomas langurs encounter (i.e., it only attacks from a relatively close distance and it is, in general, more dangerous on the ground than in trees), will affect the anti-predator behavior of Thomas langurs. Their arboreal habits may be the first protection against their predators. In addition, early detection of predators should be important. Indeed, Thomas langurs scan for predators, as they are more vigilant in the lower strata of the forest and when they have no close neighbors (Steenbeek *et al.* 1999). Moreover, larger groups are expected to be more successful at spotting predators, yet no such an effect has been found for the Thomas langurs (van Schaik *et al.* 1983). Once a predator is spotted, group-living animals may alert each other to the predator by alarm calls. Such calls are indeed given by adult male and female Thomas langurs. Moreover, the whole group was observed to mob predators (personal observation). Thus, Thomas langurs employ several anti-predation strategies.

In this chapter, the effect of predation risk on the foraging behavior of Thomas langurs is examined. A risk of predation is mainly expected in the lower strata, that is, on the ground and between 0 m and 10 m. It is assumed that the langurs are free to rest and groom in the higher strata in order to minimize the predation risk, whereas food requirements may force animals to use the lower strata. This can only be the case when the lower strata offer food items different from the higher strata. Similarly, it is expected that individuals will seek neighbors when they have to use the lower strata. In addition, individuals in a group may differ in their behavior. Adult males may be less vulnerable to predation than females because of their larger canines (Sterck 1997). Moreover, high-ranking females may reduce the risk at the expense of low-ranking ones and females with dependent offspring may be

Table 5.1. *The number of focal data of adult group members*

Group	male	Focal minutes	female	Focal minutes
B	ba	2956	bh	1985
			ma	3463
			su	2932
H	to	2658	bh	2345
			ma	631
			su	624
			ju	2078
J	da	3429	la	3314
			ol	3492
			pu	2426
M	ad	4077	ko	5334
			pa	5287
			sa	5113

more vulnerable, either themselves or their offspring, than females without. When particular individuals are more at risk, this may affect their diet, as they are less able to exploit dangerous locations.

Methods

Study site and animals

The research was conducted at the Ketambe Research Station, Gunung Leuser National Park, Leuser Ecosystem, Sumatra, Indonesia. The research area consists of primary lowland rainforest. All natural predators were present: that is, reticulated pythons, golden cats (*Felis temmincki*), clouded leopards and tigers. The langurs sometimes made alarm calls in response to small raptors (Steenbeek *et al.* 1999; Sterck 1996).

Four habituated bisexual groups of Thomas langurs were followed (Table 5.1). The groups consisted of one adult male and several females and their offspring (Sterck 1995). Groups differed in size from three to five adults and three to 11 group members. Adult females in two groups (groups J and M) could be assigned dominance ranks in food sources, whereas female dominance ranks were shared in group B and H (Sterck 1995; Sterck and Steenbeek 1997).

The female reproductive stages were divided into: female with infant of 0–3 months; female with infant of 4–12 months; and female with infant older than 12 months or no infant.

Thomas langurs are folivore–frugivores (Sterck 1995). Leaves and fruit form the major portion of diet, which also includes flowers, bark, leaf stalks, ants, birds, bird eggs, algae, soil and snails.

Observation methods

Groups of Thomas langurs were followed from dawn to dusk, usually for five days in a row. The animals were recognized individually by natural markings and differences in the size and shape of their tail and head characteristics. Focal animal samples (Martin and Bateson 1986) were collected on the adult male and the adult females of group B, H, J and M from 1989 to 1991 (Table 5.1; see also Sterck 1995 for more information on the observations). The author, Jacqueline van Oijen and Deanne Radema collected the data.

Data were collected using the instantaneous focal animal method (Martin and Bateson 1986). Originally, the day was arbitrarily divided into four periods of equal duration (before 9.00 a.m.; 9.00–12.00 a.m.; 12.00–15.00 p.m.; after 15.00 p.m.) and during each period data on one individual were collected. Because it was sometimes difficult to find the targeted individual, subsequent behavioral samples consisted of observations lasting at most 15 minutes. All samples consisting of two or more minutes were included. The samples of one individual had to be at least 30 minutes apart. The number of observation minutes was equally distributed through the day.

The behavior recorded was: feeding, resting, moving and social behavior (grooming, aggression and sex). When feeding, the food item was recorded: fruit, including seeds; leaves, encompassing young leaves, mature leaves, leaf stalks, ferns and flowers; snails; other items, including ants, bird's eggs, bird fledglings, soil, mushrooms, algae and bark; unidentified items. The height at which the animal was observed was also recorded: that is, on the ground, in the understorey (0–10 m), together forming the lower strata and the higher strata (above 10 m). Furthermore, the presence or absence of any neighbors within 5 m was recorded.

In addition, the time of entering and leaving of a food patch of each individual and the food item consumed was documented by trained local field assistants. As many patches as possible were included, but when patches were used simultaneously, only the one in which the first animal entered was included.

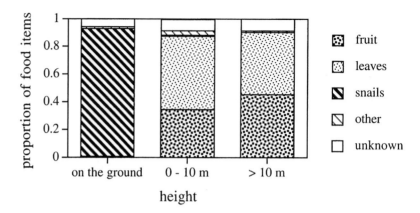

Fig. 5.1. The proportion of time that the average adult group member consumed fruit, leaves, snails, other food items or unknown food items when they were eating on the ground, between 0 m and 10 m or above 10 m.

Data analysis and statistics

The data were processed in foxBASE and EXCEL. Data were collected throughout the study years and seasonal differences in food availability do not affect the activity budget of Thomas langurs (Sterck 1995). Therefore, all available data could be combined. The diet was calculated as the proportion of time feeding on a food item of the total time feeding at that height. The activity was calculated as the proportion of time feeding or resting from the total time an individual was observed at that height. The time with neighbors was the proportion of time with neighbors of the total time in that activity at that height.

Statistical analyses were conducted in SPSS and StatView. All tests are nonparametric, two-tailed and α is 0.05. In general, individuals were compared with themselves at different heights or different reproductive stages, that is, the variables were dependent. Only for the comparison between the sexes was it assumed that the data were independent. When a Friedman two way analysis of variance (ANOVA) was significant, multiple comparisons between treatments were conducted to determine which groups differed significantly (Siegel and Castellan 1988).

Results

Diet at different heights

The different heights in the forest may offer different food. Indeed, the data show that Thomas langurs fed more often on snails while on the ground, whereas they ate mainly fruit and leaves between 0 m and 10 m or above 10 m (Fig. 5.1) (Table 5.2). In addition, *ad libitum* observations showed that, besides snails, they also ate soil and algae on the ground. Snails were found in small rivulets in the

Table 5.2. *The differences in diet, activities and number of neighbors on the ground, between 0 m and 10 m and above 10 m height were tested with a Friedman two-way ANOVA. When the Friedman test was significant, multiple comparisons between treatments were conducted to determine which groups differed significantly*

	x^2	P	N	Z_{adj}	ground vs. 0–10 m	0–10 m vs. >10 m	ground vs. >10 m
Diet							
fruit	24.57	<0.0001	14	12.67	−15.96[a]	−10.08	26.04[a]
leaves	22.29	<0.0001	14	12.67	−23.94[a]	5.88	18.06[a]
snails	17.71	0.0001	14	12.67	14.00[a]	7.948	−21.98[a]
other	4.43	0.11	14	—	—	—	—
not known	7.00	0.01	14	12.67	−10.92	−2.1	13.02[a]
Activity							
feeding	24.57	<0.0001	14	12.67	15.96[a]	10.08	−26.04[a]
resting	22.62	<0.0001	13	12.21	−14.95[a]	−9.1	24.05[a]
Neighbors while feeding							
All individuals	9.00	0.01	14	12.67	11.90	3.08	−14.98[a]
>10 min[b]	11.14	0.004	7	8.96	9.03[a]	2.94	−11.97[a]

Notes:

[a] Denotes a significant result ($p < 0.05$).

[b] Only individuals with at least 10 minutes of focal time while feeding on the ground were included.

forest. Soil was usually found between the roots of toppled trees and sometimes also from termite nests in dead trees (i.e., not on the ground). In addition, the members of groups U and B pulled algae from the Alas or Ketambe River. Only for soil and snails are food patch data available and only for snails are focal animal data available.

Thus, the food found on the ground differs from that found in the trees and most of these food items are not available in the trees. Therefore, feeding requirements may force the Thomas langurs to use the ground.

Avoidance of the ground and the understory

The ground and the understory were hypothesized to be dangerous, because of an increased predation risk. Indeed, when the langurs went to the ground, they were very cautious. Before reaching the ground they descended a little at a time and scanned their

Table 5.3. *Average patch residence time (PRT) on the ground and the number of observations (n) in patches with soil and with snails*

group	individual	PRT soil (s) [n]	PRT snails (s) [n]
B	ba	102 [3]	808 [2]
	bh	55 [1]	2717 [3]
	ma	146 [3]	1192 [5]
	su	—	2852 [3]
J	da	—	5588 [1]
	la	—	7115 [1]
	ol	—	7177 [1]
	pu	—	1368 [1]
M	ad	36 [1]	3815 [3]
	ko	35 [1]	3445 [4]
	pa	—	2547 [5]
	sa	—	2512 [5]

Notes:

—, no data.

environment thoroughly before descending a bit further. Once on the ground to feed on soil, they collected a piece and ascended quickly to consume it in the trees. Their actual time on the ground was very short (Table 5.3). This is in contrast to what is found while eating snails, which could last for an hour or longer (Table 5.3). Their behavior indicates that they were wary while feeding on snails on the ground. A slight cough, that is a mild alarm signal, sent them into the trees. Altogether, the langurs behave as if the ground is dangerous. Therefore, they were expected to avoid the ground and the understory.

In general, the langurs were on the ground during a minor part of the observations (Fig. 5.2) and were not on the ground at all during most focal days. When they were on the ground they were feeding more often than when between 0 m and 10 m or above 10 m (Fig. 5.3) (Table 5.2). Similarly, they hardly ever rested on the ground and rested much more often when between 0 m and 10 m or above 10 m (Fig. 5.3) (Table 5.2). In line with these results, they were more often feeding than resting when on the ground (Wilcoxon test: $n=13$, $Z=2.62$, $p=0.009$), whereas they were more often resting than feeding when between 0 m and 10 m (Wilcoxon test: $n=17$, $Z=3.57$, $p=0.0004$) or above 10 m (Wilcoxon test: $n=17$, Z

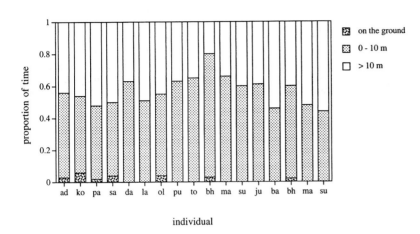

Fig. 5.2. The proportion of time that adult Thomas langurs were found at different heights.

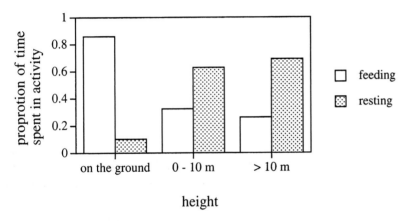

Fig. 5.3. The proportion of focal time at a particular height that the average adult group member was feeding or resting.

= 3.62, $p = 0.0003$). Thus, when they were able to avoid the ground, such as, while resting, they did avoid it. This was not found for the understory.

Height and neighbors

When neighbors provide safety, animals are expected to have neighbors more often at the more dangerous locations. Only the feeding data were used. Indeed, when they were feeding on the ground Thomas langurs had neighbors significantly more often than when above 10 m, yet the number of neighbors on the ground did not differ significantly from that between 0 m and 10 m (Table 5.2).

These measures, however, also encompass individuals that were rarely on the ground. Therefore, the same comparison was done for individuals that were feeding at least 10 focal minutes on the ground (Fig. 5.4). As expected, these animals had neighbors more often when feeding on the ground than when feeding between 0 m and 10 m or above 10 m (Table 5.2).

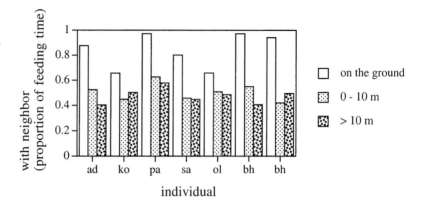

Fig. 5.4. Proportion of time with neighbors within 5 m when feeding on the ground, between 0 m and 10 m or above 10 m. Only observations of individuals feeding more than 10 focal minutes on the ground are included.

Differences between males and females

Males may be less prone to predation than females, because of their larger canines. Indeed, males had neighbors less often than females Mann–Whitney U test: $n_1=4$, $n_2=13$, $Z=2.94$, $p=0.003$). Males, however, were not more often found on the ground or between 0 m and 10 m than females (MWU test: on the ground: $n_1=4$, $n_2=13$, $Z=0.45$, $p=0.65$; between 0 m and 10 m: $n_1=4$, $n_2=13$, $Z=0.62$, $p=0.57$).

Snails seemed to be more important for females than for males. The data indicate that females fed more often on snails than males (Tables 5.3 and 5.4) and typically entered snail patches before the male (Table 5.4).

When males entered later or did not enter a rivulet, they often sat at a rather low height (below 5 m) above the females, while scanning the environment. It gave the strong impression that the males were standing guard when the females were feeding on the ground.

Differences among females
Dominance rank

Females seemed to have a rather strict order of entering the rivulets (Table 5.4). The data on group M indicate that low-ranking females (sa and ko) entered before the highest-ranking female (pa). This, however, was not corroborated by the scant data on group J.

When on the ground, the time with neighbors did not differ for high- and low-ranking females (Spearman rank correlation: $n=12$, $Z_{ties}=0.17$, $p=0.87$).

Reproductive stage

When the lower strata are indeed more dangerous, it can be expected that females with vulnerable offspring use these heights

Table 5.4. *Order of entering rivulets to feed on snails*
(1 is first, 3 or 4 is last individual to enter. When two individuals enter at the same time, the ranking was shared (e.g., 1.5). The females of group J and M are ordered according to dominance rank (Sterck and Steenbeek 1997). The females with infants are marked: [a] female with 1–3 months old infant; [b] female with 4–12 months old infant)

	males	females			
group B	ba	bh	ma	su	
	—	2[b]	3	1	
	—	2[b]	3	1	
	—	—[b]	1	—	
	4	2[b]	3	1	
	4	1[b]	3[b]	2	
group H	to	bh	ma	su	ju
	—	1.5[b]	3[b]	1.5	—[b]
group J	da	la	ol	pu	
	3	2	1	4	
group M	ad	pa	ko	sa	
	3	—	1.5	1.5	
	4	3	2	1	
	3	4	1	2	
	1.5	4	3	1.5	
	4	3	2	1	
	—	3	1	2[a]	
	—	2	1	3[a]	

less often than females without an infant (Fig. 5.5). Indeed, females with a young infant were found in the lower strata significantly less often than females without an infant (Table 5.5), whereas females with an infant from 4 to 12 months had intermediate values. Females with an infant might be expected to be surrounded by neighbors more often than females without an infant. Such an effect, however, was not found (Table 5.5).

Females with young infants did not seem to cease entering patches with snails, although when sa of group M had a small infant she entered the rivulets later (Table 5.4). In line with this

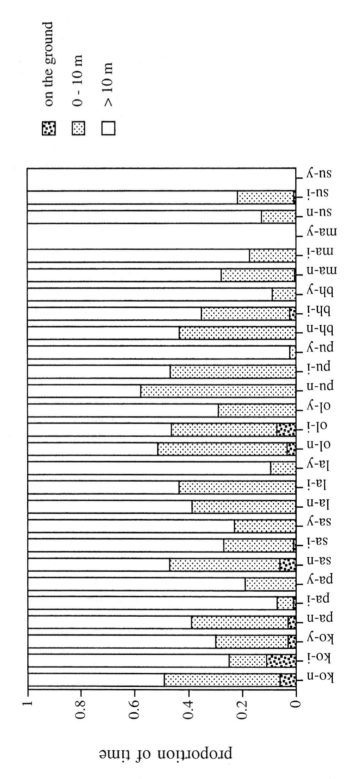

Fig. 5.5. The proportion of time that a female was found at different height categories. For each female data, when available, are presented when she had a 1–3 months old infant (y); a 4–12 months old infant (i); and an older (>12 months) or no infant (n).

Table 5.5. *The differences in use of height for females in different reproductive stages were tested with a Friedman two-way ANOVA. When the Friedman test was significant, multiple comparisons between treatments were conducted to determine which groups differed significantly*

	x^2	P	N	Z_{adj}	infant 0–3 mo vs. infant 4–12 mo	infant 0–3 mo vs. no infant	infant 4–12 mo vs. no infant
Presence at height							
On the ground	7.79	0.02	7	8.96	−8.54	9.52[a]	−0.98
0–10 m	8.86	0.01	7	8.96	−3.99	10.99[a]	−7.00
Above 10 m	8.86	0.01	7	8.96	3.99	−10.99[a]	7.00
Time with neighbors							
On the ground	3.00	0.22	2				
0–10 m	0.75	0.69	8				
Above 10 m	1.75	0.42	8				
Diet							
Fruit	0.86	0.65	7				
Leaves	5.43	0.07	7				
Snails	1.93	0.38	7				
Other	3.71	0.16	7				
Unknown	5.43	0.07	7				

Notes:
[a] Denotes a significant result ($p < 0.05$).

suggestion, the diet of females did not differ significantly with their different reproductive stages (Table 5.5).

Discussion

The effect of predation risk on foraging behavior was investigated in the Thomas langur. In this langur the lower strata of the forest are expected to represent the highest predation risk. First was established whether Thomas langurs use these risky locations more often when feeding than during resting. Next, the strategies to reduce the risk of predation were determined. Last, the consequences of avoiding particular areas on the foraging behavior were specified.

Foraging on the ground
Thomas langurs obtained food items on the ground, such as snails, algae and soil, that they could not obtain in the trees. Most time on

the ground was spent eating snails. These food items were, in the light of the amount of time spent on them, obviously important. This was also indicated by the behavior of group B. In their normal home range, this group did not have access to rivulets with snails and they visited the area across the Ketambe River (jumping between trees) to feed on snails in a rivulet on the other side, or they entered the home range of group M to feed there in a rivulet on snails. It is unclear what they obtain from snails (R. Steenbeek, personal communication). The langurs did not seem to consume the snails, but to suck them.

If the predation risk is indeed higher on the ground and in the understory, it is expected that Thomas langurs would avoid these strata when possible and only use them when they are forced to do so, such as when their feeding habits require it. Indeed, when on the ground the Thomas langurs were more often feeding than resting, while they were more often resting than feeding when in the trees. The langurs did not behave as if the trees between 0 m and 10 m were more dangerous than above 10 m. These results indicate that Thomas langurs preferred to use the trees, but that the acquisition of food items that can only be found on the ground forced them to use the ground.

Antipredation strategies

If Thomas langurs take a higher risk of predation when they feed on the ground, this should be reflected in their behavior. Several strategies are possible. In general, individuals behave differently when they are in dangerous locations. Moreover, individuals that live in larger groups may feed in dangerous places more often than those in smaller groups (e.g., wedge-capped capuchin monkeys (*Cebus olivaceus*): de Ruiter 1986). Alternatively, high-ranking individuals may reduce their predation risk at the expense of low-ranking group members (e.g., brown capuchin monkeys (*Cebus apella*): Janson 1990a,b; long-tailed macaques (*Macaca fascicularis*): van Schaik and van Noordwijk 1988). Lastly, individuals may adjust their behavior to their own risk of predation. Indeed, a number of antipredator strategies were found.

Precautions when using the ground

Thomas langurs scanned markedly frequently while descending and reacted very quickly to even mild alarm calls by climbing in trees. In general, they scanned more often when they were below 10 m than between 10 m and 20 m (Steenbeek *et al.* 1999). Thus, early detection of predators seemed an important part of their antipredation strategies.

Thomas langurs fed on the ground, but seemed to reduce their time on the ground when possible. The patch residence time was short in soil patches and the one piece of soil they consumed was carried into the trees. In contrast, the patch residence time in snail patches was very long. They fed on many snails, so carrying one up into the trees would force them to descend soon afterwards to quench their appetite for snails. Thus, feeding on snails forced them to use the ground for extended periods of time. Here they employed a different strategy and had neighbors more often than in the trees.

The male seemed to play a special role in snail patches. He did not enter the patches as often as females, but seemed to stand guard while they banqueted on the snails.

Safety through group size

Thomas langurs living in a large group do not use the ground more often than those living in a small group (Sterck 1995). Moreover, a larger group size does not improve the detection distance of potential dangers (van Schaik et al. 1983). Thus, group size in itself does not seem to provide protection. This result is expected when the density of neighbors does not vary with group size (Treves 1998) and, indeed, this was found in Thomas langurs (Sterck 1995). Therefore, group size did not seem to provide protection against predation.

Safety through dominance rank

On average, high-ranking Thomas langur females do not have more neighbors or neighbors more often than low-ranking females (Sterck 1995). They also do not use the ground more often (Sterck 1995). Yet some benefits of a high dominance rank may be present in Thomas langurs, as high-ranking females seem to enter feeding patches on the ground later than low-ranking females and thus may take less risk than other females. They may also obtain the safer places, as the rate of displacements is higher in these patches than in fruit or leaf patches (Fig. 5.4d, Sterck and Steenbeek 1997). Such an effect, however, was not found. This lack of result may be due to the limited size of the data set for this analysis. However, it is in line with the general lack of an effect of dominance rank on Thomas langur behavior and reproduction (Sterck 1995; Sterck and Steenbeek 1997).

Individual adjustments to predation risk

Although it was predicted that males had a lower predation risk than females, they did not use the ground more often than females. Males, however, had neighbors less often than females. This may be due to a lower predation risk or the females may have had more juvenile neighbors than males.

Females with a young infant were expected to have a higher predation risk than females without an infant. Indeed, females with a young infant avoided the ground. However, they did not have neighbors more often than females without an infant. Also, their diet was not affected. However, this lack of result may be due to the limited data set available for behavior on the ground. In line with the results on diet, females with a young infant did not seem to avoid snail patches, although the results hint that they did not enter first and thus may take less risk than other females.

Predator sensitive foraging in Thomas langurs

Predator sensitive foraging implies that foraging behavior is influenced by predation risk. In Thomas langurs, this seems to be the case. When they were feeding on the ground they had more neighbors and this resulted in more displacements. This seems to indicate that competition for food was increased in these patches. In general, the relative rate of aggression was high in small patches (Sterck and Steenbeek 1997). The size of snail patches, however, could not explain this high rate of displacements, because these patches were along rivulets and thus very large. The high rate was better explained by efforts to remain close together. Thus, competition not for food items but for safe locations caused the high rate of displacements in snail patches. This interpretation, however, was not corroborated by the data on individual differences of time with neighbors.

The predation risk was probably high for females with a young infant. Although predation risk affects their behavior when on the ground, it does not affect the food that Thomas langurs obtain. Therefore, their foraging behavior and diet were actually little affected by variation in predation risk.

In conclusion, Thomas langurs behaved as if the ground was dangerous and they only went there to obtain food that was not available in the trees. Although the aggregation with neighbors indicates that Thomas langurs benefit from living in a group in comparison to living solitarily, their antipredation strategies did not seem to benefit from variation in group size or differential benefits of group living. Thus, although differences in predation risk affected their behavior, no individual gained benefits that were unattainable for others.

Conclusions

1. Thomas langurs more often ate snails on the ground and fruit and leaves in the trees. Snails were a minor part of the diet, but potentially constitute an indispensable addition to the diet.

2. When on the ground, Thomas langurs were more often feeding than resting, whereas in the trees they were resting more often than feeding. Thus, when they could avoid the ground they did so.

3. Thomas langurs had neighbors more often when feeding on the ground than when feeding in trees.

4. Males were not on the ground more often than females. However, they had, on average, fewer neighbors than females.

5. High-ranking females did not have neighbors more often than low-ranking females when they were feeding on the ground.

6. Females with a young infant were less often found on the ground than females with an older or no infant. The time with neighbors and the diet did not differ for the different reproductive stages.

7. Thomas langurs showed predator sensitive behavior. They benefit from living in a group in comparison to living solitarily, but their antipredation strategies did not benefit from variation in group size or differential benefits of group living.

Acknowledgments

I thank the Indonesian Institute of Sciences (LIPI) and the Indonesian Conservation Service (PHPA) for their permission to conduct fieldwork in Indonesia. I thank Idrusman Ariga, Bahlias Putra Gayo and Suprayudin, field assistants, for their dedication to the project and their company and Jacqueline van Oijen and Deanne Radema for the collection of data. Han de Vries provided statistical advice. Comments of Lynne Miller, Eleni Nikitopoulos,

Mark Prescott, Serge Wich and an anonymous reviewer improved this manuscript. This study was financed by WOTRO, the Netherlands Foundation for the Advancement of Tropical Research. This is Leuser Management Unit Publication No. 025/2001.

REFERENCES

Alcock, J. (1998). *Animal Behavior*. 6th edition. Sunderland: Sinauer Associates, Inc.

Alexander, R.D. (1974). The evolution of social behavior. *Annual Review of Ecology and Systematics* **5**: 325–82.

de Ruiter, J. R. (1986). The influence of group size on predator scanning and foraging behaviour of wedgecapped capuchin monkeys (*Cebus olivaceus*). *Behaviour* **98**: 240–58.

Janson, C.H. (1990a). Ecological consequences of individual spatial choice in foraging groups of brown capuchin monkeys, *Cebus apella*. *Animal Behaviour* **40**: 922–34.

Janson, C.H. (1990b). Social correlates of individual spatial choice in foraging groups of brown capuchin monkeys, *Cebus apella*. *Animal Behaviour* **40**: 910–21.

Janson, C.H., and Goldsmith, M.L. (1995). Predicting group size in primates: foraging costs and predation risks. *Behavioral Ecology* **6**: 326–36.

Martin, P., and Bateson, P. (1986). *Measuring Behaviour*. Cambridge: Cambridge University Press.

Muckenhirn, N.A. (1972). Leaf eaters and their predators in Ceylon: ecological roles of grey langurs, *Presbytis entellus* and leopards. Ph.D. thesis, University of Maryland, College Park.

Siegel, S., and Castellan, Jr., N.J. (1988). *Nonparametric Statistics*, 2nd edn. New York: McCraw-Hill Book Company.

Sommer, V. (1985). Weibliche und Männliche Reproductionsstrategien der Hanuman Languren (*Presbytis entellus*) von Jodhpur, Rajastan/Indiën. PhD thesis, Universität Göttingen, Göttingen.

Stanford, C.B. (1991). *The Capped Langur in Bangladesh: Behavioral Ecology and Reproductive Tactics*. New York: Karger.

Steenbeek, R. Piek, R.C., van Buul, M., and van Hooff, J.A.R.A.M. (1999). Vigilance in wild Thomas's langurs (*Presbytis thomasi*): the importance of infanticide risk. *Behavioral Ecology and Sociobiology* **45**: 137–50

Sterck, E.H.M. (1995). Females, foods and fights. A socioecological comparison of the sympatric Thomas langur and long-tailed macaque. Ph.D. thesis Utrecht University, Utrecht.

Sterck, E.H.M. (1996). The langurs of Gunung Leuser National Park. In: C.P. van Schaik and J. Supriatna, eds., *Leuser. A Sumatran Sanctuary*, Jakarta: Yayasan Bina Sains Hayati Indonesia Depok, pp. 280–93.

Sterck, E.H.M. (1997). Determinants of female dispersal in Thomas langurs. *American Journal of Primatology* **42**: 179–98.

Sterck, E.H.M., and Steenbeek, R. (1997). Female dominance relationships and food competition in the sympatric Thomas langur and long-tailed macaque. *Behaviour* **134**: 749–74.

Treves, A. (1998). The influence of group size and neighbours on vigilance in two species of arboreal monkeys. *Behaviour* **135**: 453–81.

van Schaik, C.P. (1983). Why are diurnal primates living in groups? *Behaviour* **87**: 120–44.

van Schaik, C.P., and Hörstermann, M. (1994). Predation risk and the number of adult males in a primate group: a comparative test. *Behavioral Ecology and Sociobiology* **35**: 261–72.

van Schaik, C.P., and van Noordwijk, M.A. (1988). Scramble and contest in feeding competition among female long-tailed macaques (*Macaca fascicularis*). *Behaviour* **105**: 77–98.

van Schaik, C.P., van Noordwijk, M.A., Warsono, B., and Sutriono, E. (1983). Party size and early detection of predators in Sumatran forest primates. *Primates* **24**: 211–21.

Yoshiba, K. (1968). Local and intertroop variability in ecology and social behavior of common Indian langurs. In: P. C. Jay, ed., *Primates: Studies in Adaptation and Variability*. New York: Holt, Rinehart and Winston, pp. 217–42.

Part II • Social variables

6 • The role of group size in predator sensitive foraging decisions for wedge-capped capuchin monkeys (*Cebus olivaceus*)

LYNNE E. MILLER

Introduction

Primatologists have only recently begun to investigate the impact of predation on foraging strategies, and the many variables that might mediate this relationship. Traditionally, success in foraging and success in avoiding predators were treated as two independent phenomena. Both might depend upon group size, but the interaction was rarely considered (e.g., for capuchins, Miller 1992, de Ruiter 1986, Srikosamatara 1987). This study explores the extent to which group size influences perceived risk of predation and how this, in turn, affects food intake. Data come from an ongoing study of one population of wedge-capped capuchins (*Cebus olivaceus*), living at the Hato Piñero Biological Reserve in central Venezuela.

Previous studies with this population of capuchins revealed a strong link between group size and seasonality in food intake. Adult females living in larger groups maintain an essentially constant level of food intake throughout the year (approximately 1800 cm³ of food per female per day, on average). In contrast, those in smaller groups experience dramatic annual variation, consuming very little during the dry season (approx. 1100 cm³) but much more in the wet season (approx. 2900 cm³) (Miller 1996). This pattern is only partly the result of differences in foraging effort. During the wet season, small-group females devote significantly more time to feeding and foraging than do large-group females (68% and 55%, respectively, of their daily activity budgets) and so the disparity in food intake is predictable. However, during the dry season, foraging time is not significantly different (58% in larger groups and 54% in smaller groups) (Miller 1996). It would appear, then, that females in smaller groups receive fairly poor returns on their dry-season foraging efforts.[1]

This pattern has been attributed to seasonal fluctuations in food abundance and the effects of intergroup competition. During the dry season, food is scarce. A smaller group may expend energy looking for food but is often displaced from feeding sites by larger groups. Thus small-group foraging success is low at this time of year. During the wet season, however, food is abundant. When a smaller group is displaced from one fruit tree, another is easily found, and so these individuals can forage more successfully. In fact, members of smaller groups compensate for dry season short-falls by dramatically increasing foraging effort in the wet season, when their additional effort will pay off.

The work presented here explores an additional factor contributing to the observed pattern of food intake: the effect of predation risk on feeding ecology. Specifically, during the dry season, do larger groups concentrate their foraging efforts on resource-rich areas while smaller groups are confined to resource-poor areas, and is that difference caused by a disparity in the monkeys' perceived level of risk?

It has long been asserted that members of larger groups experience less predation because of the selfish herd effect (Hamilton 1971), the added vigilance of many eyes (e.g., Pulliam 1973), and the improved success of mobbing (e.g., Altmann and Altmann 1970). Depending on the spatial dispersion of important resources, differential predation risk may lead to disparate foraging strategies for members of larger and smaller groups. While food in low-risk areas may be accessible to both large and small groups, food in high-risk areas may be more accessible to individuals in larger groups. Combining these variables provides a working hypothesis:

Members of smaller groups are at greater risk of predation, and are therefore less willing to forage in risky areas. When scarce resources are located in risky areas, members of smaller groups sacrifice access to food in order to reduce exposure to predators. However, members of larger groups are at lower risk of predation and therefore proceed to forage in risky areas. Thus they achieve a feeding advantage over their conspecifics in smaller groups when resources are scarce.

Predatory attacks on primates are rarely observed (cf., Cheney

---

[1] This entire pattern is strongly supported by unpublished data from additional years of study (January to December, 1995, with two larger groups and five smaller groups, and January to May, 1999, with the original two study groups).

and Wrangham 1987). This may be due to several factors, including low rates of predation and the impact of human observers on the behavior of the predators themselves (e.g., Isbell and Young 1993). In the course of nearly four years of research at Hato Piñero, no predator has ever been observed capturing a capuchin monkey. However, even where actual predation is infrequent, prey animals still engage in antipredator behavior. Because predation certainly results in death, the selection to avoid it must be very strong. Antipredator behavior is therefore expected to be maintained in a population even though the incidence of predation is relatively low.

At this research site (Hato Piñero, Venezuela), there are few aerial predators large enough to capture a monkey that is in the trees, but various animals might prey on monkeys on the ground. There are several species of predatory cats, including jaguar (*Panthera onca*), puma (*Felis concolor*), and ocelot (*Felis pardalis*); these animals generally hunt at night but have also been observed traveling about in the daytime. There are venomous and constricting snakes, such as rattlesnakes (*Crotalus* spp.) and boas (*Boa constrictor*). Around pools and streams there are abundant caiman crocodile (*Caiman crocodilus*). There are no data to indicate that rates of predation at Hato Piñero actually *are* higher for monkeys on the ground, but these subjects certainly *behave* as if they *perceive* the ground to be a risky area. Monkeys on the ground alarm call more frequently and respond to alarm calls more readily (by quickly climbing trees) than they do while in the trees (personal observation); they also scan at higher rates (personal observation; see also Srikosamatara 1987). These observations suggest that, although predation occurs rarely, its impact is still reflected in the monkeys' behavior, and this risk is manifested in heightened vigilance while on the ground.

Despite the risks, these capuchins do come to the ground during certain times of the year. In the wet season, food resources – including both ripe fruits and invertebrates – are located in the trees, and tree hollows provide water. In the dry season, however, important resources are located on the ground. A terrestrial bromeliad (*Bromelia pinquin*) is the second most commonly eaten fruit by these monkeys and, during three months of the dry season, accounts for approximately 30% of food intake (by volume) (Miller 1998). In addition, invertebrate species exploited by capuchins are generally in their larval and pupal stages during the dry season and are therefore less readily accessible to these monkeys (Miller 1992). Protein intake at this time of year comes primarily in the form of large apple snails (*Pomacea* spp.), which also represent a large proportion of the dry season diet (Miller 1992; see also Robinson 1986). Furthermore, tree hollows dry out during these months, forcing thirsty monkeys to make use of ponds or other terrestrial sources of water. Thus, while the monkeys can make a living in the trees throughout the wet season, the scarcity of food in the dry season (cf., Miller 1992, 1996) makes terrestrial resources attractive.

Building upon this set of information, this study seeks to test the following:

Prediction 1: During the dry season, adult females in the larger group spend more time on the ground than do those in the smaller group.

Prediction 2: During the dry season, adult females in the larger group acquire more food from the ground than do those in the smaller group.

Methods

Subjects:

The subjects of this study belonged to a natural population of wedge-capped capuchins (*Cebus olivaceus*). This is a medium sized platyrrhine, with males weighing approximately 3 kg, females about 2.5 kg (Robinson and Janson 1987, Rowe 1996). They rely heavily on ripe fruits and invertebrate material, supplementing their diet with flowers, new leaves and the occasional vertebrate prey item (Robinson and Janson 1987). At this site, fruit feeding accounted for about 17% of the daily activity budget (DAB), but provided about 65% of food intake by volume; invertebrate foraging required over 40% of the DAB and provided about 35% of food intake by volume (on average) (Miller 1997).

Quantitative data were collected for the adult female members of two focal groups. The larger group (LG) comprised between 34 and 40 individuals, of which approximately 12 were adult females. The smaller group (SG) included 14 to 16 members, of which six were adult females (for details on group composition, see Miller 1992). Adult females were the focus of data collection. While this methodological choice limits comparison across age–sex classes, it allowed for more rapid accumulation of data on this one class and thus produced a more detailed profile of their feeding ecology. Furthermore, the fitness of females, more than of males, is influenced by foraging success (cf., Trivers 1972).

Both study groups were fully habituated to human presence when data collection began, so that the observer could approach to within 3 m without having an obvious impact on a subject's behavior, even when that subject was on the ground. Habituation was a fairly easy task as this population has been protected for nearly 50 years. Both groups used the entirety of the study site, and thus had the same resources and predators present in their home ranges.

Site

The research took place at the Hato Piñero Biological Reserve, located in the *llanos* (plains) of Venezuela. This region is characterized by pronounced seasonality, with heavy rains from May through October and little precipitation from November through April (on average; for detail, see Miller 1998). During the wet season, the region floods, from a few centimeters to a couple of meters deep, depending on soil composition and ground contour. Climatic conditions lead to fluctuations in food type and abundance, with larger volumes of fruit and greater numbers of invertebrates available to these monkeys in the wet season than in the dry (Miller 1992, 1996). The study site itself is a 270-hectare area within an extensive stretch of semideciduous dry tropical forest. A grid of trails provides approximately 45 km of transects which were used to assess forest composition, tree density, and fruit abundance and dispersion (Miller 1992, 1998).

Data Collection and Analysis

Preliminary work (e.g., cutting and mapping trails, habituating and identifying subjects) was carried out from April, 1989, through May, 1990. Behavioral observations reported here were made from July, 1990, through June, 1991. Data were collected on an opportunistic basis, throughout the 12-hour period of daylight, whenever contact with the study groups could be made. The protocol required focal animal samples of adult females, of 30-second duration, at one-half-hour intervals. Information relevant to this study was the subject's location (e.g., whether on the ground or in the trees) and the species of food being eaten (for details on additional data collected, see Miller 1992). For this analysis, a sample was considered to demonstrate use of the ground if the subject was actually on the ground during the sample, or the subject was eating a food item that could only be harvested from the ground (i.e., a bromeliad fruit or an apple snail). The data set does not allow evaluation of each female's behavior, and so it is impossible to assess

Fig. 6.1. Variations with season and group size in use of the ground. See text for statistical values.

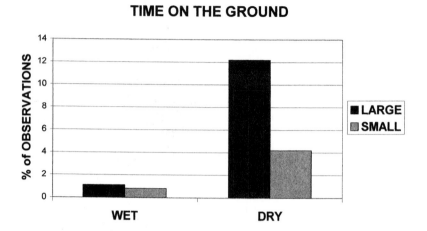

TIME ON THE GROUND

individual variation. Instead, data were pooled for each group so that final figures represent (for Prediction 1) the percentage of total observations (per group) in which the ground was used and (for Prediction 2) the percentage of total feeding observations in which a terrestrial food was used. The chi-squared test was used to evaluate the strength of the relationship between group size and use of the ground, and between group size and food intake, stratified by season.

Results

Prediction 1 was supported by the data (Fig. 6.1). To begin, there was a clear seasonal disparity in use of the ground. Females in both study groups made use of the ground statistically significantly more often in the dry season than in the wet season ($\chi^2 = 103.03$, $p < 0.001$). However, during the dry season, females in the larger group made use of the ground significantly more frequently than did those in the smaller group (12% of observations for LG, 4% for SG; $\chi^2 = 23.39$, $p < 0.001$).

Prediction 2 was also supported by the data (Fig. 6.2). Again, there was a seasonal disparity in the number of observations in which food was obtained from the ground, with females in both study groups exploiting terrestrial resources significantly more frequently in the dry season than in the wet ($\chi^2 = 59.13$, $p < 0.001$). More importantly, during the dry season, adult females in the larger group ate terrestrial foods significantly more often than did those in the smaller group (for LG, 17.4% of feeding observations involved bromeliad fruits or apple snails, for SG, 8.4%; $\chi^2 = 23.39$, $p < 0.001$).

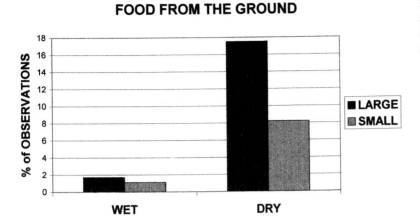

Fig. 6.2. Variations with season and group size in food acquired from the ground. See text for statistical values.

Discussion

Scientists studying the behavioral ecology of capuchin monkeys have paid considerable attention to the relationship between group size and measures of reproductive success (e.g., de Ruiter 1986, Janson 1985, 1988, Robinson 1988, Srikosamatara 1987). Where large groups are shown to have an advantage, it is generally supposed that this is through the mechanism of reduced mortality due to predation, or of increased foraging success. This study suggests that these two factors may interact to bestow fitness advantages.

The results of this study clearly support Prediction 1. During the dry season, females in the larger group came to the ground far more frequently than did those in the smaller group. This observation is consistent with the hypothesis that members of larger groups perceive themselves to be less vulnerable to terrestrial predators. However, in the context of predator sensitive foraging, it is inadequate simply to demonstrate that certain individuals make use of a potentially risky area; it must also be clear that these individuals gain a feeding advantage by this behavior. The data pertaining to Prediction 2 show that females in the larger group received a greater proportion of their food from the ground than did those in the smaller group. Extrapolating from the available data provides a more tangible sense of the disparity: On average, a large group female procured about 315 cm^3 of food per day from the ground (17.4% of feeding observations $\times$ 1800 cm^3 of food consumed per day); a small-group female found only 90 cm^3 of food per day in this potentially risky area (8.4% of 1100 cm^3). Taken together, these results support the hypothesis that membership in a larger group enhances access to resources in risky areas, and increases foraging opportunities.[2]

Alternative explanations for this observed pattern of behavior must be considered before predator sensitive foraging is accepted as the principal factor. For example, intergroup differences in feeding are often caused by disparities in microhabitat (e.g., for capuchins, Brown and Zunino 1990, Chapman and Fedigan 1990, Miller 1998) or may be part of learned traditions (e.g., McGrew 1983, Miller 1998). However, these explanations are unlikely in this case because the two groups in question are part of the same population and have completely overlapping home ranges (in this study site). As individuals move between groups, group-level differences in behavior should diminish, except where those differences are adaptive.

Another possibility is that members of larger groups are *forced* to exploit terrestrial resources because of higher intragroup feeding competition. That is, perhaps larger groups cannot be supported by the food supplies in the trees, and thus they must come to the ground despite the risks. Perhaps large-group females are displaced from arboreal feeding sites by larger-bodied males. However, this interpretation is inconsistent with the previously described pattern of food intake. If large-group females were using terrestrial resources out of sheer necessity, while small-group females were fed adequately in the trees, we would expect to see lower food intake for those in the larger group. However, the opposite is true. During the dry season, a female in the large group eats approximately 70% more, by volume, than does one in the small group (on average; Miller 1996). Presumably females in larger groups could subsist through the dry season on the limited amount of food eaten by those in smaller groups. The fact that females in larger groups actually eat more than those in smaller groups indicates that large groups forage on the ground not because they have to but because they can.

Returning to the explanation conventionally provided for group-level differences in feeding success, perhaps larger groups simply monopolize terrestrial resources, and thus smaller groups are excluded from these feeding sites not by predation risk but by intergroup competition. Surely intergroup competition is an important factor in the feeding ecology of this population. Smaller

---

[2] Data from 1995 (pooled for seven groups) and 1999 (returning to the original large and small groups) yield similar results; however, the samples are small and therefore analyses are of limited value. However, taken as a whole, this data set produces a strikingly consistent pattern.

groups are often displaced from arboreal food patches when larger groups come into view (e.g., out of intimidation) but do not hesitate to feed there when alone. However, in the case of terrestrial foods, displacement is not the only factor. Even when no other group is around (i.e., no other group has been detected by the observer throughout the hours of data collection), members of smaller groups still hesitate to exploit rich bromeliad patches. This suggests that a combination of influences serves to shape the feeding behavior of these animals.

Data collected over several years of study support a relationship between group size and predator sensitive foraging: Ground foraging is more common in larger groups and appears to result in greater access to important resources. The weakest part of this model is the assertion that terrestrial foraging is risky. There are no data from this study site to indicate that predation is more common on the ground than in the trees, or that larger groups actually experience lower mortality rates (cf., Hill and Dunbar 1998). However, the subjects' behavior, such as more frequent alarm calling and scanning (cf., de Ruiter 1986, Srikosamatara 1987), suggests that they perceive the ground to be a dangerous place. Furthermore, these subjects have often been seen mobbing potential ground predators, such as snakes and small cats (personal observation); other studies of capuchins have yielded similar observations (e.g., Boinski 1988, Defler 1980). Additional subjective data from the Hato Piñero capuchins further support these assertions. For example, all monkeys show considerable hesitation before approaching the ground. When monkeys wish to harvest bromeliad fruits, they usually come down just long enough to grab a bunch of fruits and then quickly return to the trees to eat; rarely do they stay on or even near the ground. When monkeys approach a pond or stream to drink, they almost always do so in groups of two or three, huddled closely together, often with their tails entwined as they lean over the water. Occasionally, individuals may travel on the ground for relatively long distances (e.g., 100 m), but this has been observed almost exclusively for members of larger groups and for adult males; females with infants virtually never come to the ground. All of these observations serve to support – if only circumstantially – the general hypothesis presented here.

The strength in numbers may allow members of larger groups to forage in risky areas and to displace smaller groups from feeding sites. Combined, these factors evidently provide large-group females with the means to maintain constant levels of food intake throughout the year, despite seasonal changes in resource abundance. In contrast, those in smaller groups experience much lower intake during the dry season. Although they may compensate for dry season shortfalls by foraging more (and more successfully) during the wet season, the fluctuation itself may have fitness ramifications. Long-term demographic data from a neighboring population of wedge-capped capuchins reveal a positive correlation between group size and reproductive success (Robinson 1988). Given that gestation usually occurs during the dry season, reduced foraging success for small-group females at this time of year may be the mechanism responsible for this trend. Thus, the small-group foraging pattern comes at a cost. However, we must conclude that this strategy represents an adaptive compromise, a way to maximize fitness within the constraints of small-group membership. Differential risk of predation therefore represents one important variable in the development of foraging strategies for members of large and small groups.

Acknowledgments

First and foremost, I want to thank Sr. Antonio Julio Branger for his continuing support of my research at Hato Piñero. *Muchas gracias* to the staff and community at Hato P. for their kindness, and to Robert Harding for introducing me to Hato P. I am grateful to those who have helped with this manuscript: Drs S.A. Miller, P.A. Garber, L.A. Isbell and D.P. Mullin. Portions of this research were supported by grants from the Center for Field Research (Earthwatch) and I am grateful to the many volunteers who helped me find and follow monkeys.

REFERENCES

Altmann, S.A., and Altmann, J. (1970). *Baboon Ecology*. Chicago: University of Chicago Press.

Boinski, S. (1988). Use of a club by a white-faced capuchin (*Cebus capucinus*) to attack a venomous snake (*Bothrops asper*). *American Journal of Primatology* **14**: 177–80.

Brown, A.D., and Zunino, G.E. (1990). Dietary variability in *Cebus apella* in extreme habitats: Evidence for adaptability. *Folia Primatologica* **54**: 187–95.

Chapman, C.A., and Fedigan, L.M. (1990). Dietary differences between neighboring *Cebus capucinus* groups: Local traditions, food availability or responses to food profitability? *Folia Primatologica* **54**: 177–86.

Cheney, D.L., and Wrangham, R.W. (1987). Predation. In: B.B. Smuts, D.L. Cheney, R.L. Seyfarth, R.W. Wrangham and T.T. Struhsaker, eds. *Primate Societies*. Chicago: University of Chicago Press, pp. 227–39.

de Ruiter, J. (1986). The influence of group size on predator scanning and foraging behaviour of wedge-capped capuchin monkeys (*Cebus olivaceus*). *Behaviour* **98**: 240–58.

Defler, T.R. (1980). Notes on interactions between the tayra (*Eira barbara*) and the white-fronted capuchin (*Cebus albifrons*). *Journal of Mammalogy* **61**(1): 156.

Hamilton, W.D. 1971. Geometry for a selfish herd. *Journal of Theoretical Biology* **31**: 295–311.

Hill, R.A., and Dunbar, R.I.M. (1998). An evaluation of the roles of predation rate and predation risk as selective pressures on primate grouping behaviour. *Behaviour* **135**: 411–30.

Isbell L.A., and Young, T.P. (1993). Human presence reduces predation in a free-ranging vervet monkey population in Kenya. *Animal Behaviour* **45**: 1233–5.

Janson, C.H. (1985) Aggressive competition and indivivdual food consumption in wild brown capuchin monkeys (*Cebus apella*). *Behavioral Ecology and Sociobiology* **18**: 125–38.

Janson, C.H. (1988). Food competition in brown capuchin monkeys (*Cebus apella*): Quantitative effects of group size and tree productivity. *Behaviour* **105**: 53–76.

McGrew, W.C. (1983). Animal foods in the diets of wild chimpanzees (*Pan troglodytes*): Why cross-cultural variation? *Journal of Ethology* **1**: 46–61.

Miller, L.E. (1992). *Socioecology of the Wedge-Capped Capuchin Monkey* (Cebus olivaceus). Ph.D. Dissertation. University of California, Davis.

Miller, L.E. (1996). The behavioral ecology of wedge-capped capuchin monkeys (*Cebus olivaceus*). In: M.A. Norconk, A.L. Rosenberger, P.A. Garber, eds. *Adaptive Radiations of Neotropical Primates*. New York: Plenum Press, pp. 271–88.

Miller, L.E. (1997). Methods of assessing dietary intake: A case study from wedge-capped capuchins in Venezuela. *Neotropical Primates* **5**(4): 104–8.

Miller, L.E. (1998). Dietary choices in *Cebus olivaceus*: A comparison of data from Hato Piñero and Hato Masaguaral. *Primate Conservation* **18**: 42–50.

Pulliam, H.R. (1973). On the advantages of flocking. *Journal of Theoretical Biology* **38**: 419–22.

Robinson, J.G. (1986). Seasonal variation in use of time and space by the wedge-capped capuchin monkeys, *Cebus olivaceus*: Implications for foraging theory. *Smithsonian Contributions to Zoology* **431**: 1–60.

Robinson, J.G. (1988). Group size in wedge-capped capuchin monkeys *Cebus olivaceus* and the reproductive success of males and females. *Behavioral Ecology and Sociobiology* **23**: 187–97.

Robinson, J.G., and Janson, C.H. (1987). Capuchins, squirrel monkeys and atelines: socioecological convergence with Old World primates. In: B.B. Smuts, D.L. Cheney, R.L. Seyfarth, R.W. Wrangham and T.T. Struhsaker, eds. *Primate Societies*. Chicago: University of Chicago Press, pp. 69–82.

Rowe, N. (1996). *The Pictorial Guide to the Living Primates*. New York: Pogonias Press.

Srikosamatara, S. (1987). Group Size in Wedge-Capped Capuchin Monkeys (*Cebus olivaceus*): Vulnerability to Predators, Intragroup and Intergroup Feeding Competition. Ph.D. Dissertation. University of Florida, Gainesville.

Trivers, R.L. (1972). Parental investment and sexual selection. In: B. Campbell, ed. *Sexual Selection and the Descent of Man 1871–1971*. Chicago: Aldine, pp. 136–79.

7 • Group size effects on predation sensitive foraging in wild ring-tailed lemurs (*Lemur catta*)

MICHELLE L. SAUTHER

Introduction

Models that seek to address the costs and benefits of social group-living often posit resource competition as one disadvantage of social living (Alexander 1974, Janson and Goldsmith 1993, Terborgh and Janson 1986, van Schaik 1983, van Schaik and van Hooff 1983, Wrangham, 1980). Such models differ in the importance of inter- versus intragroup competition, and on what factors encourage individuals to tolerate living in larger groups. 'Predation–intragroup feeding competition' models (PFC) (Alexander 1974, Terborgh and Janson 1986, van Schaik 1983, van Schaik and van Hooff 1983) all focus on intragroup feeding competition as the ultimate cost for group living, which intensifies as group size increases. For such models, increased predator detection and hence a reduction in predator risk is the driving force behind larger groups, and group size fluctuates as a response to these two factors. Indeed, van Schaik (1983: p. 138) has posited, 'group living is disadvantageous with respect to feeding and that the avoidance of predation confers the only universal selective advantage of group living in diurnal primates.' There have been few direct tests of the PFC model for primates, but larger groups of wild long-tailed macaques were better at predator detection (van Schaik *et al.* 1983) and smaller groups of yellow baboons did exhibit behaviors more conducive to monitoring predators or predator escape (Stacey 1986). It is clear, however, that there is no simple relationship between predation and group size (Boinski and Chapman 1995, Treves 1999). As such, there is a need for more specific quantitative analyses to determine the relationship between group size and predator pressure. Most studies which have tested the PFC model have focused on overall patterns for many groups within the same area (e.g., van Schaik and van Noordwijk 1988). While this can provide general support or rejection, it does not allow more specific socio-ecological comparisons. A different approach is taken here. In this chapter I look, in detail, at the relationship between group size and antipredator behaviors for two differently sized groups of wild ring-tailed lemurs, *Lemur catta*. Specifically I will explore how variable rates of predator contact affect predator sensitive foraging relative to group size.

Study site and species

Research was conducted at the Beza Mahafaly Special Reserve, which is located approximately 35 km northeast of the town of Betioky (23° 30′ S lat., 44° 40′ E long.), Madagascar. During the 12-month study (November 1987 to November 1988), I concentrated on ring-tailed lemurs living within an 80-ha fenced and guarded portion of the reserve that is part of a larger, continuous tract of dry gallery forest. The forest contains mostly deciduous and semi-deciduous trees (Sauther 1992), and is characterized by more lush vegetation near the Sakamena River that becomes more xerophytic vegetation as one moves away from the river (see Fig. 7.1). Grazing by sheep and cattle is prohibited, and a rich understory of herbs and lianas is present during the wet season. The reserve contains a wealth of birds, mammals, reptiles and insects that are representative of southwestern Madagascar. It also includes a natural complement of predators, some of which prey on lemurs within the reserve (Ratsirarson 1985, Sauther 1989). The habitat is highly seasonal with approximately 99% of the annual rainfall occurring between November and April (Sauther 1999). There is a hot/wet season, a cool/dry season and a transitional period during which temperatures and rainfall gradually increase. This seasonality in rainfall affects food availability, with more fruits and leaves available during the wet season (Sauther 1999).

Ring-tailed lemurs are semi-terrestrial, diurnal prosimian primates. They live in large, female-dominant social groups in which females have linear hierarchies (Jolly 1966, Sauther 1992, Taylor 1986). Ring-tailed lemurs can best be described as opportunistic foragers as they exploit seasonal resources as they become available, utilizing a wide variety of foods including leaves, flowers, fruit and insects (Sauther 1993, 1999). Group members are highly synchronous in their behavior, with each group foraging and feeding as one unit. Along with a seasonal environment, ring-tailed lemurs at Beza Mahafaly exhibit a highly seasonal reproductive cycle, with females coming into estrus during May and giving birth during

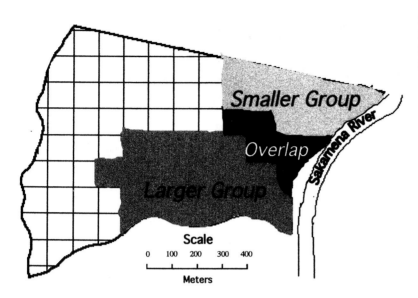

Fig. 7.1. Home ranges for the larger and smaller ring-tailed lemur groups at Beza Mahafaly Madagascar.

October (Sauther 1991). Infants are weaned around 5 months of age, and females produce infants on a yearly basis (Sauther *et al.* 1999).

Ring-tailed lemur predation

Ring-tailed lemurs are semi-terrestrial, and at Beza Mahafaly the percentage of time spent on the ground varies by month ($\bar{x} = 32\%$, range = 3%–75%). As such, the lemurs are exposed to a number of terrestrial as well as arboreal predators (Jolly 1966, Sussman 1977, Sauther 1989). At Beza Mahafaly these include the Malagasy giant hognose snake, *Leioheterodon madagascariensis*, boa constrictors, *Boa mandtria* (formerly *Boa madagascariensis*), the fossa, *Cryptoprocta fossa*, Indian civet, *Viverricula indica*, domesticated dogs and cats, and two raptors, the Madagascar harrier hawk, *Polyboroides radiatus* and the Madagascar buzzard, *Buteo brachypterus* (Goodman *et al.* 1993, Sauther 1989). A number of predation events have been documented at Beza Mahafaly. An infant ring-tailed lemur was observed being predated on by the Madagascar harrier hawk (Ratsirarson 1985), and bones of adult and juvenile ring-tailed lemurs have been found below roosting sites of this raptor at another site (Goodman *et al.* 1993). An infant ring-tailed lemur was also observed being attacked and killed by an Indian civet and there is evidence of lemur predation by feral cats and/or fossa within the reserve (Goodman *et al.* 1993, Sauther 1989). Antipredator defense in this species involves vigilance and mobbing behaviors as well as alarm calls, once potential predators are sighted (Jolly 1966, Sussman 1972, Sauther 1989). Field and captive work indicates that representational signaling is present in ring-tailed lemur alarm calls (Macedonia 1990, Sauther 1989), but may be absent in the alarm calls of some other lemur species (Macedonia 1990). It should be noted that previous assumptions that Madagascar's lemurs are more immune to predation due to the absence of large felids (van Schaik and van Hooff 1983) ignores the impact of important carnivores such as the fossa, and aerial raptors which are known predators of young lemurs (Goodman *et al.* 1993, Ratsirarson 1985). Many researchers have noted the importance of predation pressure on lemur behavior (Jolly 1966, Macedonia 1993, Richard 1978, Sauther 1989, Sussman 1972, Goodman, *et al.* 1993). There is thus no evidence to indicate that lemurs are under relaxed predation pressure, and if predation plays an important role in affecting foraging behavior then its impact should be measurable in populations of ring-tailed lemurs.

Foraging location

Predation behaviors by the predators themselves are an important component in understanding how their presence affects primate predation sensitive foraging such as foraging location. The harrier hawk, endemic to Madagascar (Dee 1986), is a large raptor, 60–62 cm in body length (Milon *et al.* 1973). Given its large size, it is not easily able to move within closed environments such as closed canopy forests (Sauther, personal observation). Observations of the behavior of the harrier hawk at Beza Mahafaly indicate that it especially favors open environments where terrestrial prey may be attacked from above. These hawks were commonly observed in fields, and would prey on chickens in open areas of local villages by swooping down from above (Sauther, personal observation). In this study, all encounters with the harrier hawk occurred in more open terrain such as meadows or clearings, with the hawk either flying above or perched in a dead standing tree. The Madagascar buzzard is a smaller raptor, 48–51 cm (Milon *et al.* 1973) and it is more adept at maneuvering in closed environments than the harrier hawk (Pidgeon, personal communication). It is also commonly found perched in trees and tends to take its prey on the ground either from the air or from a stationary perch (Brown and Amadon 1968). At Beza Mahafaly these hawks were encountered in closed canopy forest. At another research site, Berenty, buzzards have been observed swooping at ring-tailed lemur troops carrying infants

(Jolly, personal communication). Little has been said of predation by snakes on lemurs, but at Beza Mahafaly ring-tailed lemurs gave predator calls, and leaped into bushes and trees when the large Malagasy giant hognose snake and boas were encountered on the ground. Both snakes are avoided by the lemurs (Sauther 1989). From the lemur's point of view, some foraging locations are potentially riskier than others. Terrestrial foraging is expected to be an especially vulnerable area because this is where the large snakes were encountered and because the harrier hawk appears to favor ground attacks in open areas. However, a number of important ring-tailed lemur foods, including herbaceous leaves, are found primarily at ground level (Sauther 1999). In this study, I test the following predictions regarding foraging location:

1. When predation risk is high ring-tailed lemurs should avoid riskier foraging areas, specifically terrestrial foraging.
2. Smaller groups should avoid these areas more than larger groups.
3. Groups avoiding these areas should have reduced foraging/feeding measures.

Mixed-species association

Mixed-species associations among primates may provide a number of benefits, including increased access to food resources and greater safety from predators (Waser 1986). Ring-tailed lemurs at Beza Mahafaly periodically form associations with the other large-bodied lemur in the reserve, Verreaux's sifaka, *Propithecus verreauxi*. These two lemurs react strongly to each other's alarm calls (Sauther 1989) which suggests that such associations may be formed as a response to predator pressure. I predict that:

1. When predator risk is high (more predator contact or presence of vulnerable infants), ring-tailed lemurs should form mixed-species associations with sifaka.
2. Smaller ring-tailed lemur groups should join with sifaka more often than larger groups.
3. Lemurs in associations should exhibit reduced foraging ability.

Group cohesion during foraging

Numerous studies have suggested that predator pressure correlates with greater cohesion (e.g., Caine 1993, van Schaik and van Noordwijk 1986a,b). Additionally, both wild and captive studies indicate that cohesion leads to increased feeding competition (Barton 1989, Furuichi 1983, Janson 1985, Mori 1977, Robinson 1981). In this study I predict that:

1. Greater contact with predators should be correlated with greater cohesion (closer distance between nearest neighbors).
2. Smaller groups should show this pattern more than larger groups.
3. Cohesion should be correlated with reduced foraging success.

Use of new hectares

Because new or less frequented areas are likely to contain unknown risks, and because smaller groups may have a reduced ability to detect predators (e.g., van Schaik *et al.* 1983), we may expect a group size effect on foraging in new areas. Ring-tailed lemurs do not visit all parts of their home ranges equally, and some areas are entered at much lower frequencies than others are. As such, these areas may be more dangerous especially since a major predator, the harrier hawk, is a 'sit and wait' ambush predator. Three predictions are made regarding this for ring-tailed lemurs:

1. Use of new hectares should be correlated with higher contact with predators.
2. Use of new hectares should be correlated with greater access to resources.
3. Smaller groups should use new hectares less frequently than larger groups.

Materials and methods

Study subjects

Group size for troops living within the reserve ranged from nine to 22 individuals at the beginning of the study (Sussman 1991). While I wanted to address how group size might affect the lemur's ecological patterns, I also needed to compare adjacent groups. This was in order to avoid confounding the analysis by comparing groups in dramatically different parts of the reserve (e.g., the drier portions versus the wetter areas). Of all groups with adjacent home ranges, I chose the two groups that were the most different in terms of size. Group size varied during the year through the addition and deletion of transferring males and infant births and deaths. The larger group ranged from 14 to 16 members while the smaller group ranged from eight to nine members. Given the in-depth nature of the study, only two groups could be compared. Each adult and sub-adult individual within these groups has been collared and tagged as part of a long-term study of ring-tailed lemur demography (see Sussman 1991). Thus the data presented here are based on individually identifiable primates. Troops were habituated to observers, allowing close range observations (1–2 m). Using the focal animal sampling method (Altmann 1974), each adult and subadult individual's behavior was recorded at 5-minute intervals 1 day per month, with two individuals within the same group being followed by an assistant and myself on the same day. Specific feeding data is based on monthly frequencies relative to the number of focal individuals followed for that month. For example, if the larger group fed on leaves 100 times during a month, this was averaged by the number of focal individuals in the group (e.g., $16 \div 100 = 0.16$).

Foraging location

In this study, feeding was strictly defined as actual ingestion of food items and does not include handling time. Handling of and search for food items is defined as foraging. Foraging locations were recorded following the tree quadrat method (see Fig. 7.2) (Mendel 1976). For this method, a tree is divided into thirds along the vertical and horizontal axes and each of these quadrats is numbered sequentially, 1–9. If the animal forages on the ground this is assigned number 10. Such a method is advantageous as it is independent of height and thus can be used for all forests, regardless of their architecture (Garber and Sussman 1984).

Mixed-species associations

During the study year I observed mixed-species associations of ring-tailed lemurs and sifaka on 34 occasions. Throughout the year each incident was noted, and the context of such associations was recorded as feeding/foraging, resting, and predation (e.g., either a predator was observed or the two species responded to each other's predator alarm calls during the association) (Sauther 1989).

Group cohesion during foraging

At 15-minute intervals the behavior and location of the focal animal's nearest neighbor were recorded. These data were used to generate nearest neighbor distances in order to evaluate whether predator pressure best explains cohesion or noncohesion during

Fig. 7.2. Foraging locations used by ring-tailed lemurs.

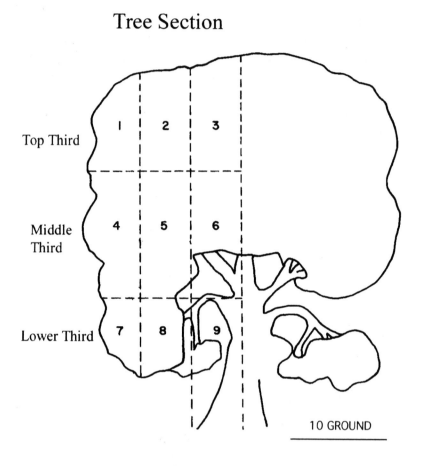

Tree Section

Top Third

1 | 2 | 3

Middle Third

4 | 5 | 6

Lower Third

7 | 8 | 9

10 GROUND

feeding. Data are presented only for nearest neighbors who were also feeding or foraging. Individuals with nearest neighbor 8 m or more away were scored as 'solitary feeding.' Individuals who had nearest neighbors less than 3 m away were scored as 'group feeding.'

Use of new hectares

Home ranges of the two groups are shown in Fig. 7.1. One square hectare paths have been cut throughout this area, and trees along the paths are color coded so it was possible to generate detailed information regarding home and day ranges. Each group's location was mapped every 15 minutes and these data were used to determine ranging patterns. The normal home range of the larger group was greater than that of the smaller group (larger = 30 ha, smaller = 19 ha). The two groups had home ranges adjacent to one another, with an 8-ha overlap. As a measure of foraging effort, I also recorded the total number of new hectares entered during each study month by each group. Because a group could enter the same hectare many times during the day, 'new hectares' refers only to the total number of hectares entered once. Thus, if a group entered a total of ten hectares, but five of these were entered twice, the number of new hectares would be only five.

Predation

For this study predator pressure was estimated by recording all encounters with potential diurnal predators throughout the year. Although it can be argued that the presence of an observer may reduce overall encounters, the *pattern* of encounters should not change. This can provide one measure of predation pressure, given that predation is rarely directly observed in wild populations.

Statistics

A common limitation in the study of animal behavior is that often the same individuals are sampled repeatedly, and thus the samples lack both randomness and independence. Furthermore, sample sizes are often relatively small. Because of this, nonparametric methods are often used to analyze such data (Thomas 1986). However, many nonparametric methods lack efficiency, and information is lost (Thomas 1986), especially if the data set contains interval data. Furthermore, nonparametric tests make the specific assumption that the sample being analyzed was produced through random sampling which, as noted above, is often not a valid assumption in behavioral studies (Sussman *et al.* 1979). A more powerful method for the analysis of such data is the use of randomization tests, which results in a distribution-free test (Edgington 1980, Manly 1991). The concept of such tests is very old, but before the advent of computers such an approach was not practical. In this study, a t-test with systematic data permutation was used to test for statistically significant differences between the two groups (Manly 1991, see Sauther 1992). I employed 1000 random permutations to generate a t statistic, but when results were nearly significant I repeated the analysis with 5000 random permutations to guard against type II errors (Manly 1991). Standard correlation coefficients were also generated for some data sets and analyzed using the Statview II application (Abacus Concepts, Berkeley, California). For this study significance was set at $p < 0.05$.

Table 7.1. *Effects of predator pressure on ring-tailed lemur foraging/feeding locations*

Group	Foraging level in tree/ground	High predator pressure (Percentage)	Low predator pressure (Percentage)	t	P
Small ring-tailed lemur group	Top third	13.59	8.41	1.89	0.09
	Middle third	11.98	5.74	2.76	0.02
	Lower third	2.56	3.15	0.50	0.63
	Ground	5.50	19.59	2.47	0.03
Large ring-tailed lemur group	Top third	11.44	7.78	1.54	0.16
	Middle third	7.77	8.47	0.33	0.75
	Lower third	3.64	2.31	2.56	0.04
	Ground	16.47	18.65	0.37	0.72

Results

Foraging/feeding location

Predation sensitive foraging might be expressed in terms of foraging locations. To test for this, each group was compared for foraging location relative to predator pressure. Monthly predator encounters between 0 and 3 were judged as 'low,' encounters greater than 3 were deemed 'high.' The smaller group preferentially foraged within the middle level of trees when predator pressure was high (see Table 7.1). They also avoided foraging on the ground when pressure was high, but readily foraged at this level when predator pressure was low. In the larger group, members preferred foraging at the lower level of feeding trees during periods of high predator pressure, but there was no predator effect on their use of the ground level. Thus, for the larger group, the level of predator pressure had little effect on their foraging behavior, whereas in the smaller group members varied their foraging and feeding locations relative to potential predation threat. I compared the two groups for the frequency of intake of fruits and leaves at ground level and in the trees relative to predator pressure to determine if avoidance of feeding locations negatively impacted feeding behavior. There was no difference for terrestrial fruit feeding, which is not surprising as most fruit feeding occurs in the trees. However, the smaller group had a significantly lower intake of leaves at ground level when predator pressure was high (mean monthly intake during low predation pressure$=1.93$; mean monthly intake during high predation pressure$=0.50$; $t=2.40$, $p=0.04$). The smaller group also showed less fruit intake when feeding in trees during periods of high predator pressure (mean monthly intake during low predation pressure$=11.33$; mean monthly intake during high predation pressure$=4.95$; $t=2.08$, $p=0.04$). There were no such effects for the larger group relative to foraging locations, predator pressure and food intake.

Mixed-species associations

Ring-tailed lemurs might also form mixed-species associations to increase predator detection during foraging. Of the 34 mixed-species associations noted during the study, 15% ($n=5$) occurred during resting, 26% ($n=9$) occurred during predator encounters, and the majority, 59% ($n=20$), occurred when both species were feeding/foraging. To test for group size effects I ran correlations between mixed-species associations and the number of predator encounters by each ring-tailed lemur group per month. Only the smaller group showed a significant association across the year (smaller group: $r=0.83$; $t=4.65$, $p=0.0009$). These associations were not random but showed peaks during periods of infant lemur vulnerability, specifically during weaning and birth seasons (Sauther, 1989) (see Fig. 7.3). To assess the effects of such associations on feeding behavior I ran correlations for total time spent feeding, fruit, leaf and flower feeding and solitary and group feeding. Of these, mixed-species associations were negatively correlated with flower feeding ($r=0.57$, $n=12$, $t=-2.20$, $p=0.05$) and positively correlated with close feeding ($r=0.74$, $n=12$, $t=3.43$, $p=0.006$).

Group cohesion during foraging

Predator pressure (measured here as the number of predators encountered per month by each group) should be negatively correlated with time spent alone, and positively correlated with close nearest neighbors. In other words, as more predators are encountered, solitary behaviors should decrease, and more individuals should maintain closer proximity to one another. Correlations indicate that neither 'solitary' nor 'group' nearest neighbor measures were significantly associated with predator encounters in either group. However, the smaller group did behave differently than the larger group when entering new hectares. The number of new hectares entered and the 'group' nearest neighbor pattern

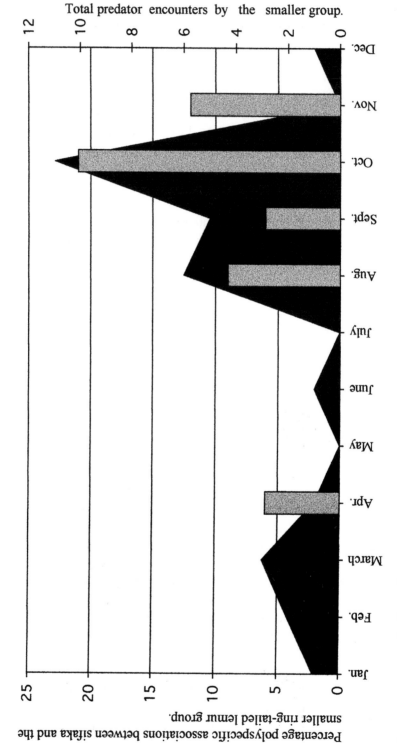

Fig. 7.3. Percentages of polyspecific associations between members of the small *Lemur catta* group and *Propithecus verreauxi* (bars) relative to total monthly predator encounters by the small group (areas). *Propithecus verreauxi* weaning occurs from January to March (Richard *et al.* 2000), *Lemur catta* weaning occurs from February to March (Sauther 1989). *Propithecus verreauxi* births occur during June and August (Richard *et al.* 2000) and *Lemur catta* births occur during October (Sauther 1989).

were positively correlated ($r=0.51$, $n=12$, $t=2.05$, $p=0.049$). Unlike the smaller group, the larger group spread out when entering new hectares ($r=0.72$, $n=12$, $t=3.31$, $p=0.008$).

Use of new hectares

For the larger group the number of new hectares entered was correlated with a higher number of predator encounters ($r=0.62$, $n=12$, $t=2.53$, $p=0.03$), but this was not seen in the smaller group. The larger group also entered a greater number of new hectares during foraging than the smaller group ($t=2.60$, $p<0.03$). While there were obvious disadvantages to entering new hectares, such behavior might also result in the group encountering more potential resources. I therefore ran correlations between new hectares entered and fruit and leaf feeding and time spent foraging. New hectares were positively correlated with fruit feeding (new hectares and fruit feeding: $r=0.75$, $n=12$, $t=3.58$, $p=0.005$) but not with leaf feeding or foraging. In other words, by entering more new hectares the larger group spent more time feeding on fruits, but also significantly increased their encounters with predators.

Discussion

It is to be expected that numerous constraints may mediate predation sensitive foraging in primates. In this study, the smaller group did not exhibit increased cohesion when more predators where encountered, as might be expected. However, the smaller group differed from the larger group in that members maintained closer proximity to one another when entering new hectares. As new areas likely contain unknown risks, and smaller groups may have a reduced ability to detect predators (e.g., van Schaik *et al.* 1983), such behavior by the smaller group does support predation sensitive foraging. In the larger group, members were actually more spread out when entering new hectares, suggesting that there may be a group size effect. Whether this is due to increased predator detection or increased feeding competition or a combination of both is unclear. For the larger group, entering more new hectares paid off in greater fruit feeding, but it was also potentially costly in that they encountered more predators.

There was support for predation sensitive foraging with regards to where ring-tailed lemurs foraged for food. The smaller group avoided foraging on the ground and spent more time within the middle third of trees when predator pressure was high. The smaller group thus exhibited changes in foraging behavior that could lessen attacks by these predators, especially by decreasing terrestrial foraging when predator pressure was high. In contrast, the larger group showed fewer antipredator behaviors than the smaller group (e.g., predator pressure had no effect on foraging on the ground) suggesting again that larger groups may be able to 'afford' to take more foraging chances than smaller groups. There were potential costs associated with avoidance of particular foraging levels. Terrestrial herbs and leaves from ground vines are extremely important to ring-tailed lemurs at Beza Mahafaly, with 48% of their entire leaf diet coming from plant resources located on the ground. By avoiding the ground when predator pressure was high, the smaller group reduced their leaf intake. Foraging more in the middle third of feeding trees may also have reduced fruit feeding ability for the smaller group. In comparison, foraging location had no effect on resource use in the larger group.

Intergroup encounters between ring-tailed lemur groups are often violent, and usually result in one group being chased away (Sauther *et al.* 1999) and sifaka groups within the south will engage in ritualized 'battles' (Jolly 1966, Richard 1977). Thus, forming same species associations does not appear to be an option for either ring-tailed lemurs or sifaka at Beza Mahafaly. One way to increase group size is to form mixed-species associations instead. Mixed-species associations among primates may provide a number of benefits, including increased access to food resources and greater safety from predators (Waser 1986). For ring-tailed lemurs and sifaka it is not clear how their associations would increase access to food resources. In fact, ring-tailed lemurs often displace sifaka from feeding trees (Sauther, personal observation). Furthermore, in this study mixed-species associations were associated with reduced flower feeding and increased cohesion during feeding and thus potentially greater direct contest competition for resources. These associations do appear explicable within the context of predation sensitive foraging. First, most of the mixed-species associations were in the context of feeding/foraging. Second, sifaka and ring-tailed lemurs reacted strongly to each other's alarm calls, even when the group calling was far away (Sauther 1989). Third, and perhaps most importantly, these associations peaked during periods of infant vulnerability. Sifaka at Beza Mahafaly wean their infants from January to March (Richard *et al.* 2000); ring-tailed lemurs wean their infants between February and March (Sauther 1991) (see Fig. 7.3). Weaned infants are still small, but independent and are thus extremely vulnerable. Peaks also occurred during ring-tailed lemur and sifaka birth seasons – another period of infant vulnerability, especially because ring-tailed lemur infants are precocious and begin to move off their mothers at an early age (Gould 1990). Infant mortality is high for both species at this site (sifaka annual infant mortality 33%–80%, Kubzdela 1997; ring-tailed lemur annual infant mortality 30%–50%, Sussman 1991). Because these associations were significant only for the smaller group, it suggests that this group was compensating for its smaller size by forming a mixed-species association. Such an association could also be valuable to sifaka, which characteristically exhibit smaller group size at Beza Mahafaly (most groups having four to six individuals; Kubzdela 1997, Kubzdela and Enafa 1995) than ring-tailed lemurs (groups average 13–15 individuals; Sussman 1991). Predation by the harrier hawk has been posited as the main cause of infant mortality for sifaka (Karpanty and Goodman 1999), and these hawks have also been seen attacking and even carrying off a female adult sifaka with an infant on her back at another site (Goodman et al. 1993).

Conclusion

The results presented here provide some support for the existence of predation sensitive foraging in ring-tailed lemurs. While predator pressure alone did not affect grouping patterns, it did affect foraging locations. Group size was a key factor. As predicted, the smaller group not only avoided potential 'hot spots' (e.g., terrestrial foraging) for predation by snakes, civets, fossa, and raptors when predation encounters were high, they also formed mixed-species associations at times when both species' infants were more vulnerable and predator pressure was greater. However, these behaviors had some foraging costs. Compared to the smaller group, the larger group exhibited fewer examples of predation sensitive foraging, and even engaged in behaviors that resulted in more predator encounters. What is not clear is whether such behavior reflects superior predator detection associated with larger group size, or greater feeding competition driving individuals in larger groups to 'take more chances.' There were no immediate disadvantages to such behaviors, and there were clear advantages, such as increased fruit feeding.

Contrary to prediction, there was a lack of an association

between group dispersion and predator encounters for these lemurs, which reduces support for the PFC model as a universal selective force for group-living. A similar lack of support is reported by Treves (1999) in a much more specific study. Predator encounters did not result in greater intragroup cohesion in red-tailed monkeys, but did so for red colobus in response to chimpanzee predation. Given that primates must balance antipredator responses with other maintenance and social behaviors, such responses are perhaps best viewed as a result of particular social and ecological environments. Group size thus becomes an important variable in understanding predation sensitive foraging.

Acknowledgments

My work in Madagascar would not have been possible without the generous help by B. Rakotosamimanana, B. Andriamihaja, V. Randrianasolo, J. Andriamampianina, and P. Rakotomanga, J.A. Rakotoarisoa, and A. Rakotozafy. The hospitality and assistance of the people of Analafaly and the Beza Mahafaly reserve guards is also gratefully acknowledged. I especially thank Lynne Miller for her insight and guidance in the preparation of this manuscript. Research at Beza Mahafaly was funded in part by the National Science Foundation, the National Geographic Society and the L.S.B. Leakey Fund.

REFERENCES

Alexander, R.D. (1974). The evolution of social behavior. In: R.F. Johnston, ed., *Annual Review of Ecology and Systematics*. Palo Alto: Annual Reviews Inc., pp. 325–83.

Altmann, J. (1974). Observational study of behavior: sampling methods. *Behaviour*, **49**: 227–67.

Barton, R.A. (1989). Foraging strategies, diet and competition in olive baboons. Unpublished Ph.D. Thesis, University of St. Andrews.

Boinski S., and Chapman, C.A. (1995). Predation on primates: Where are we and what's next? *Evolutionary Anthropology* **4**: 1–3.

Brown, L., and Amadon, D. (1968). *Eagles, Hawks, and Falcons of the World*, Vol. 2. New York: McGraw-Hill Book Company.

Caine, N.G. (1993). Flexibility and cooperation as unifying themes in *Saguinus* and social organization and behaviour: The role of predation pressures. In: A.B. Rylands, ed., *Marmosets and Tamarins: Systematics, Behaviour, and Ecology*. Oxford: Oxford University Press, pp. 200–19.

Dee, T.J. (1986). *The Endemic Birds of Madagascar*. London: Chameleon Press, Limited.

Edgington, E.S. (1980). *Randomization Tests*. New York: Marcel Dekker Inc.

Furuichi, T. (1983). Interindividual distance and influence of dominance on feeding in a natural Japanese macaque troop. *Primates* **24**: 445–55.

Garber, P.A., and Sussman, R.W. (1984). Ecological distinctions between sympatric species of *Saguinus* and *Sciurus*. *American Journal of Physical Anthropology*, **65**: 135–46.

Goodman, S.M. O'Connor, S., and Langrand, O. (1993). A review of predation on lemurs: implications for the evolution of social behavior in small, nocturnal primates. In: P.M. Kappeler and J.U. Ganzhorn, eds., *Lemur Social Systems and their Ecological Basis*. New York: Plenum, pp. 51–66

Gould, L. (1990). The social development of free-ranging infant *Lemur catta* at Berenty reserve, Madagascar. *International Journal of Primatology*, **11**: 297–318.

Isbell, L.A., and Young, T.P. (1993). Social and ecological influences on activity budgets of vervet monkeys, and their implications for group living. *Behavioral Ecology and Sociobiology* **32**: 377–85.

Janson, C.H. (1985). Aggressive competition and individual food consumption in wild brown capuchin monkeys (*Cebus apella*). *Behavioral Ecology and Sociobiology* **18**: 125–38.

Janson, C.H., and Goldsmith, M.L. (1993). Predicting group size in primates: foraging costs and predation risks. *Behavioral Ecology* **6**: 326–36.

Jolly, A. (1966). *Lemur Behavior*. Chicago: University of Chicago Press.

Karpanty, S.M., and Goodman, S.M. (1999). Diet of the Madagascar Harrier-Hawk, *Poloboroides radiatus*, in Southeastern Madagascar. *Journal of Raptor Research* **33**: 313–16.

Kubzdela K.S. (1997). Sociodemography in diurnal primates: The effects of group size and female dominance rank on intra-group spatial distribution, feeding competition, female reproductive success, and female dispersal patterns in white sifaka, *Propithecus verreauxi verreauxi*. Ph.D. Thesis. Chicago: University of Chicago.

Kubzdela, K.S., and Enafa, E. (1995). Group size, social structure, and population dynamics in: *Propithecus v. verreauxi*. In: B.D. Patterson, S.M. Goodman., and J.L. Sedlock, eds., *Environmental Change in Madagascar*. Chicago: The Field Museum, pp. 32–3.

Macedonia, J.M. (1990). What is communicated in the antipredator calls of lemurs: Evidence from playback experiments with ring-tailed and ruffed lemurs. *Ethology* **86**: 177–90.

Macedonia, J.M. (1993). Adaptation and phylogenetic constraints in the antipredator behavior of ringtailed and ruffed lemurs. In: P.M. Kappeler and J.U. Ganzhorn, eds, *Lemur Social Systems and their Ecological Basis*. New York: Plenum, pp. 67–84.

Manly, B.F.J. (1991). *Randomization and Monte Carlo Methods in Biology*. New York: Chapman and Hall.

Mendel, F. (1976). Postural and locomotor behavior of *Alouatta palliata* on various substrates. *Folia Primatologica* **35**: 147–78.

Milon, P.H. Petter, J.J., and Randrianasolo, G. (1973). *Oiseaux Faune de Madagascar.* Tananarive: 35. ORSTOM, and Paris: CNRS.

Mori, A. (1977). Intra-troop spacing mechanisms among Japanese monkeys. *Primates* **14**: 113–59.

Ratsirarson, J. (1985). *Contribution à l'Étude Comparative de l'Éco-ethologie de Lemur catta dans Deux Habitats Différents de la Réserve Spéciale de Beza-Mahafaly.* Universite de Madagascar: Memoire de Fin d'Etudes.

Richard, A.F. (1977). The feeding behaviour of *Propithecus verreauxi.* In: T.H. Clutton-Brock, ed., *Primate Ecology: Studies of Feeding and Ranging Behavior in Lemurs, Monkeys, and Apes.* London: Academic Press, pp. 72–96.

Richard, A.F. (1978). *Behavioural Variation: Case Study of a Malagasy Lemur.* Pennsylvania: Bucknell University Press.

Richard, A.F., Dewar R.E., Schwartz M., and Ratsirarson, J. (2000). Mass change, environmental variability and female fertility in wild *Propithecus verreauxi. Journal of Human Evolution* **39**: 381–91.

Robinson, J.G. (1981). Spatial structure in foraging groups of wedge-capped capuchin monkeys *Cebus nigrivittatus. Animal Behavior* **29**: 1036–56.

Sauther, M.L. (1989). Antipredator behavior in groups of free-ranging *Lemur catta* at Beza Mahafaly Special Reserve, Madagascar. *International Journal of Primatology* **10**: 595–606.

Sauther, M.L. (1991). Reproductive behavior of free-ranging *Lemur catta* at Beza Mahafaly Special Reserve, Madagascar. *American Journal of Physical Anthropology* **84**: 463–77.

Sauther, M.L. (1992). Effect of Reproductive State, Social Rank and Group Size on Resource Use Among Free-ranging Ringtailed Lemurs (*Lemur catta*) of Madagascar, Ph.D. thesis, Washington University, St. Louis.

Sauther, M.L. (1993). Changes in the use of wild plant foods in free-ranging ringtailed lemurs during lactation and pregnancy: Some implications for early hominid foraging strategies. In: N.L. Etkin, ed., *Eating on the Wild Side: The Pharmacologic, Ecologic, and Social Implications of Using Noncultigens.* Tucson: University of Arizona Press, pp. 240–56.

Sauther, M.L. (1999). The interplay of phenology and reproduction in ringtailed lemurs: Implications for ringtailed lemur conservation. In: C.S. Harcourt, R.H. Crompton and A.T.C. Feistner, eds., *Biology and Conservation of Prosimians, Folio primatologica* **69**: (Suppl.1), 309–20.

Sauther, M.L., Sussman, R.W., and Gould, L. (1999). The Socioecology of the Ringtailed Lemur: Thirty-five years of study. *Evolutionary Anthropology* **8**: 120–32.

Stacey, P.B. (1986). Group size and foraging efficiency in yellow baboons. *Behavioral Ecology and Sociobiology* **18**: 175–87.

Sussman R.W. (1972). An ecological study of two Madagascan Primates: *Lemur fulvus rufus* (Audebert) and *Lemur catta* (Linnaeus). Unpublished Ph.D. Thesis, Durham: Duke University.

Sussman R.W. (1977). Socialization, social structure, and ecology of two sympatric species of *Lemur.* In: S. Chevalier-Skolnikoff and F.E. Poirier, eds., *Primate Bio-social Development: Biological, Social, and Ecological Determinants*. New York: Garland Publishing, Inc., pp. 515–28.

Sussman, R.W. (1991). Demography and social organization of free-ranging *Lemur catta* in the Beza Mahafaly Reserve, Madagascar. *American Journal of Physical Anthropology* **84**: 43–58.

Sussman, R.W., O'Fallon, W.M., Sussman, L., and Buettner-Janusch, J. (1979). Statistical methods for analyzing data on daily activity cycles of primates. *American Journal of Physical Anthropology* **51**: 1–14.

Taylor, L. (1986). Kinship, Dominance, and Social Organization in a Semi Free-ranging Group of Ringtailed Lemurs (*Lemur catta*). Unpublished Ph.D. dissertation, Washington University.

Terborgh, J., and Janson, C.H. (1986). The socioecology of primate groups. *Annual Review of Ecological Systematics* **17**: 111–35.

Thomas, D.H. (1986). *Refiguring Anthropology: First Principles of Probability and Statistics*. Chicago: Waveland Press, Inc.

Treves A. (1999). Has predation shaped the social systems of arboreal primates? *International Journal of Primatology* **20**: 35–67.

van Schaik, C.P. (1983). Why are diurnal primates living in groups? *Behaviour* **87**: 120–44.

van Schaik, C.P., and van Hooff, J.A.R.A.M. (1983). On the ultimate causes of primate social systems. *Behaviour* **85**: 91–117.

van Schaik, C.P., and van Noordwijk, M. (1986a). The hidden costs of sociality: Intra-group variation in feeding strategies in Sumatran long-tailed macaques (*Macaca fascicularis*). *Behaviour* **99**: 296–315.

van Schaik, C.P., and van Noordwijk, M (1986b). The evolutionary effect of the absence of felids on the social organization of the Simeulue monkey (*Macaca fascicularis fusca*). *Behavioral Ecology and Sociobiology* **13**: 173–81.

van Schaik, C.P., and van Noordwijk, M. (1988). Scramble and contest in feeding competition among female longtailed macaques (*Macaca fascicularis*). *Behaviour* **105**: 77–98.

van Schaik, C.P., van Noordwijk, M.A., Warsono, B., and Sutriono, E. (1983). Party size and early detection of predators in Sumatran forest primates. *Primates* **24**: 211–21.

Waser P.M. (1986). Interactions among primate species. In: B.B. Smuts, D.L Cheney, R.M. Seyfarth, R.W. Wrangham, and T.T. Struhsaker, eds., *Primate Societies*, Chicago: University of Chicago Press, pp. 210–26.

Wrangham, R.W. (1980). An ecological model of female-bonded primate groups. *Behaviour* **75**: 262–300.

8 • Species differences in feeding in Milne Edward's sifakas (*Propithecus diadema edwardsi*), rufus lemurs (*Eulemur fulvus rufus*), and red-bellied lemurs (*Eulemur rubriventer*) in southeastern Madagascar: Implications for predator avoidance

DEBORAH J. OVERDORFF, SUZANNE G. STRAIT & RYAN G. SELTZER

Introduction

Despite the difficulty in documenting it, predation pressure has long been acknowledged to play a powerful role in shaping the behavioral ecology of extant primates (Anderson 1986, Cheney and Wrangham 1987, Isbell 1994). Its role as a selection pressure on the lives of prosimian primates, however, has been debated (van Schaik and Kappeler 1996). As research on prosimian social structure in southern and western parts of Madagascar began to flourish in the late 1960s and 1970s, it was believed that predators had little influence on prosimian behavioral ecology (Jolly 1966, Richard 1978). This perspective, however, has been reevaluated with the dramatic increase in research on a wider range of species in a variety of habitats.

It is now clear that a variety of prosimian species exhibit a wide range of antipredatory behaviors. Goodman (1994a,b) has suggested that these behaviors are residual tactics which remain in the repertoire following the recent extinction of a large crowned eagle (*Stephanoaetus mahery*; but see Csermely 1996). However, recent long-term studies have demonstrated that predation pressure remains a real threat from extant raptors (Goodman *et al.* 1998, Gould 1996, Sauther 1989, Schwab 1999), mammalian carnivores (*Cryptoprocta ferox* and *Galidia elegans*; Overdorff and Strait 1995, Wilson *et al.* 1989, Wright 1998, Wright and Martin 1995), owls, and boa constrictors (Goodman *et al.* 1993, Rakotondravony *et al.* 1998, Wright and Martin 1995).

Very little is known, however, about which behavioral strategies arboreal prosimians implement to avoid predation (but see Macedonia and Pollock 1989, Oda 1998, Sauther 1989, Wright 1998, Wright *et al.* 1997). The purpose of this chapter therefore is to compare and contrast the habitat use and feeding patterns of the Milne Edward's sifaka (*Propithecus diadema edwardsi*), the rufus lemur (*Eulemur fulvus rufus*), and the red-bellied lemur (*Eulemur rubriventer*) to discuss whether these patterns could be considered predator avoidance tactics. All three species have similar diets and exploit fruit from many of the same plant species (Hemingway 1996, 1998, Overdorff 1993a, Overdorff and Strait 1998). The main differences between these species are in body size and group size. Sifaka are the largest prosimian at the site (adult body size 5–6 kg; Overdorff and Strait, unpublished data), and live in groups with multiple adult males and females, ranging between three to nine individuals (Overdorff *et al.* 1999, Wright 1995). Rufus lemurs also live in large groups (mean, eight individuals) but are about one-third the body size of sifaka (2 kg). In comparison, red-bellied lemurs are similar to rufus lemurs in body size (2 kg) but live in small, pair-bonded groups (three to five individuals; Overdorff 1993a,b, 1996a,b). This comparison provides an excellent opportunity to examine possible predator avoidance tactics in three species which exploit food from similar sources but differ in body size and group size.

Various studies have identified a wide range of 'risks' primates face while feeding. Areas such as the upper and lower edges of food patches and the ground leave an individual more visually exposed to either an aerial or arboreal/terrestrial predator (Eason 1989, Miller Chapter 6 this volume, Peetz *et al.* 1992, Skorupa 1989, Struhsaker and Leakey 1990). Small groups of *Cebus* for example avoid feeding on the ground more often than large groups (de Ruiter 1986, Miller Chapter 6 this volume). Small food patches can also be risky because individuals may be forced to feed alone or the number of possible nearest neighbors is limited, thereby diminishing the benefits of having 'more eyes and ears' (Baldellou and Henzi 1992, Cords 1990, Gould *et al.* 1997, Rose and Fedigan 1995, Treves 1998, van Schaik *et al.* 1983).

How might these factors influence prosimian feeding habits? Despite their larger body size, adult sifaka are susceptible to predation by *Cryptoprocta ferox* and offspring may be vulnerable to aerial predators (Dollar *et al.* 1997, Rasoloarison *et al.* 1995, Overdorff and Strait unpublished data, Wright 1998, Wright *et al.* 1997). Rufus and red-bellied lemurs which are much smaller bodied, also fall prey to *C. ferox* (Overdorff and Strait 1995) and a variety of arboreal raptors (Goodman *et al.* 1993). In this chapter, we discuss whether or not these three species show foraging adaptations which reflect predation threat. One of the problems faced when making a multi-species comparison regarding feeding habits is that body size, substrate preferences, and differences in locomotor ability can also influence feeding choices. How do we know if these feeding patterns are strategies that reflect predator avoidance? In this chapter, there is some control for this problem by comparing two congeneric species (rufus and red-bellied lemurs) that have the same body size, similar diets, habitat use patterns (Overdorff 1996a,b), and locomotor abilities (Dagosto 1995) but differ dramatically in group size. Sifaka live in groups similar in size to rufus lemurs but are larger in body size compared to the two *Eulemur* species. Yet, the diet of all three species is remarkably similar regarding the preferred plant species and parts eaten. If similar morphology and body size drives feeding habits, then red-bellied lemurs and rufus lemurs should have more similar feeding patterns compared to the sifaka. However, if increased group size provides increased safety from predation while feeding, sifaka and rufus lemurs are likely to show more similar feeding patterns when compared to red-bellied lemurs. Assuming this is the case, we predicted that: (a) sifaka and rufus lemurs will exploit foods from riskier areas than red-bellied lemurs by feeding in the outer and upper edges of food patches and feed on the ground and lower canopy more often; (b) sifaka and rufus lemurs will use smaller patches which could be more risky as they are less likely to have a nearest neighbor while feeding.

Methods

Study site and study species

The Ranomafana National Park (RNP) was established in 1991 and encompasses 41 000 ha ($21°2'-21°25'$ S, $47°18'-47°37'$ E) of lowland to montane rain forest in southeast Madagascar. This region supports ten primate species in addition to the three study species (Wright 1992). This study was conducted at the Vatoharanana study site which lies 5 km south from the main RNP research site. Monimiaceae, Cunoniaceae, Lauraceae and Myrtaceae are predominant plant families (Schatz and Malcomber unpublished data). Cooler, drier months fall between May and September (rainfall <150mm month^{-1}, range 4–20 °C) and warmer, wetter months fall between October and March (rainfall >150 mm month^{-1}, range 11–31 °C). For this paper, two study groups of sifaka (Group I, six individuals; Group II, five individuals), two groups of rufus lemurs (Group I, nine individuals; Group II, seven individuals), and three groups of red-bellied lemur (Groups I and II, three individuals; Group III, four individuals) were used to determine if there were any species differences in feeding ecology and if these differences might be shaped by predator avoidance tactics.

General methodology

Data on sifaka and red-bellied lemur diet, feeding time, canopy use, and whether the focal animal had a nearest neighbor or not were collected during three separate field seasons (May–August, 1994–96) as part of a larger project to examine the influence of diet on dental micro-wear (see Strait and Overdorff 1995). Similar data were collected on rufus lemurs from 1988–89 (Overdorff 1993a,b, 1996a,b). We describe these general patterns here and further examine differences between lemur species by comparing feeding patterns in four plant species (*Cabucala* sp, *Chrysophyllum madagascariensis*, Anacardiaceae sp, and *Ficus pyrifolia*) which were used most often by all three lemur species. All three lemur species ate fruit from these four plant species for more than 50% of the time thereby allowing a more direct comparison in some analyses of how each lemur species used similar food sources.

Study groups were located using radio telemetry equipment and a focal animal was observed by two observers daily from dawn to dusk. One observer was responsible for watching the animal and calling out observations to the second observer who acted as a recorder. Across the day, all feeding bouts greater than 10 seconds were recorded (see Overdorff 1993a,b for rationale). The beginning and ending time of each bout, the plant species, and the plant part ingested were noted. The recorder and observer would switch roles several times across the day to prevent observer fatigue. All statistical tests were conducted within each data set for each year of study. If no differences were found, data were combined to present an overall picture for each species. Nonparametric statistics were used and significance level was set at $p < 0.05$.

Canopy level and location within patches

Each patch was categorized as a ground source, lower-story, middle-story, and upper-story tree (which included emergent trees). To test whether the three lemur species fed in riskier locations (ground or

Fig. 8.1. The percentage of feeding bouts initiated by sifaka, rufus lemurs, and red-bellied lemurs in ground, lower-story, middle-story, and upper-story patches.

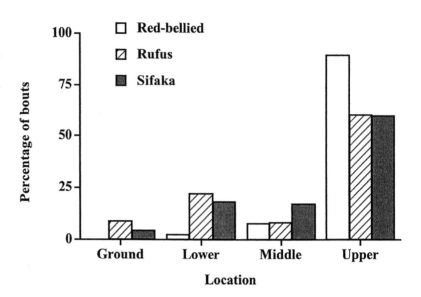

lower-story), the number of times each focal animal used each category was calculated and converted to a percentage.

Patch size and nearest neighbor
Patch size was recorded for each feeding bout for all three species and patches were classified as small, medium, and large following Overdorff (1996b). It was assumed that smaller patches (which are most often found in lower parts of the canopy) would be riskier, since fewer individuals could feed simultaneously and small patches have less dense foliage and smaller branches than large patches located higher in the forest canopy. This assumption will be examined using the nearest neighbor data.

Results

Canopy level use
As expected, sifaka and rufus lemurs fed more often on the ground, and used more lower- and middle-story trees than red-bellied lemurs which fed most often in the upper canopy ($\chi^2 = 37.86$, $p < 0.0001$, df = 6, Fig. 8.1).

Patch size and nearest neighbor
Rufus lemurs and sifaka initiated an equal number of feeding bouts in small, medium, and large patches while almost all of the red-bellied lemur feeding bouts were initiated in large patches

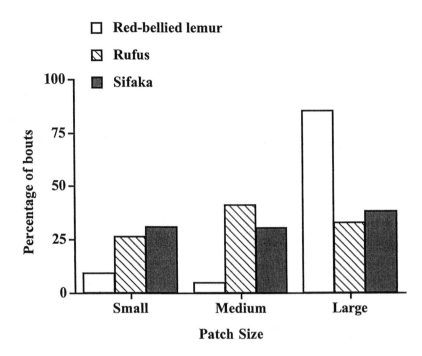

Fig. 8.2. The percentage of feeding bouts initiated by sifaka, rufus lemurs, and red-bellied lemurs in small, medium, and large patches.

($\chi^2 = 71.78$, $p < 0.0001$, df$=4$, Fig. 8.2). This pattern was consistent for four shared plant species as well (Table 8.1).

In general, sifaka and rufus lemurs spent less time feeding with a nearest neighbor in small and medium patches than in large patches ($\chi^2 = 44.09$, $p < 0.0001$, df$=4$, Fig. 8.3). The inverse relationship was observed for red-bellied lemurs: they had nearest neighbors 100% of the time in small and medium patches and spent less time feeding near another individual in large patches. This pattern was also consistent for the four shared plant species.

Discussion

In this study, sifaka, rufus lemurs, and red-bellied lemurs exhibited behaviors that are considered to be predator evasion tactics in other primate species (Eason 1989, Miller Chapter 6 this volume, Peetz *et al.* 1992, Skorupa 1989, Struhsaker and Leakey 1990). While large, upper canopy trees make up the majority of all three of these species' diets, sifaka and rufus lemurs exploited a wider range of food sources in all levels of the canopy compared to red-bellied lemurs. Not only did they feed more often in small and medium patches that are found in the lower and middle levels of the forest canopy, they fed on more ground sources (herbs, vines, and dirt).

Table 8.1. *Proportion of bouts feeding in small, medium, and large patches of the four select plant species*

| Plant species | Prosimian | Patch size | | | n | |
		small	medium	large		
Cabucala sp.	red-bellied	4.4	26.1	69.5	23	$x^2 = 52.86$
	rufus	24.1	25.9	50.0	62	$p < 0.0001$
	sifaka	45.5	15.9	38.6	44	df = 4
Ficus pyrifolia	red-bellied	0.0	0.0	100.0	12	$x^2 = 17.83$
	rufus	0.0	17.4	82.6	179	$p < 0.0001$
	sifaka	0.0	10.0	90.0	30	df = 2
Chrysophyllum madagascariensis	red-bellied	0.0	0.0	100.0	24	$x^2 = 88.71$
	rufus	13.1	37.5	49.4	275	$p < 0.0001$
	sifaka	25.0	30.0	45.0	60	df = 4
Anacardiaceae sp.	red-bellied	0.0	0.00	100.0	16	$x^2 = 131.85$
	rufus	3.5	26.3	70.1	57	$p < 0.0001$
	sifaka	4.5	34.6	60.9	23	df = 4

Fig. 8.3. Nearest neighbor patterns by patch size for sifaka, rufus lemurs, and red-bellied lemurs.

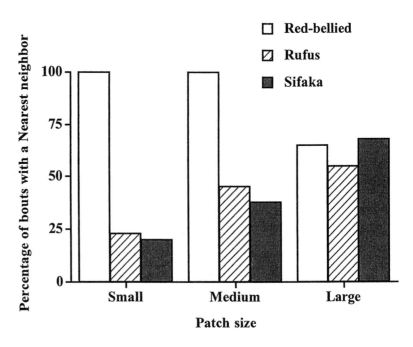

Smaller sized patches located in the lower levels of the forest could potentially be more risky because a group cannot feed together; and individuals in this study were less likely to have a nearest neighbor in small and medium patches. Sifaka and rufus lemurs were more easily startled when feeding in these types of resources and alarm called more frequently when feeding on or near the ground level (Overdorff personal observation, Wright 1998). Other *Eulemur* species that are similar in group size and/or body size to red-bellied lemurs also avoid ground sources, range higher in the canopy, and move cryptically while feeding (*E. coronatus, E. mongoz*; Curtis *et al.* 1999, Freed 1996, Rasmussen 1999, Wilson *et al.* 1989). *E. mongoz*, which live in the highly seasonal deciduous dry forests of Madagascar, will shift to an almost nocturnal activity cycle when canopy cover is most minimal (Rasmussen 1999).

There were also striking differences in feeding style between the species; red-bellied lemurs were far more cryptic while feeding than sifaka or rufus lemurs. Although all three species exploited ripe fruits from many of the same plant species, sifaka actively extracted seeds rather than consume fruit flesh. While these differences are clearly related to morphological differences (see Yamashita 1996 for dental differences), they can have important implications for how easily an animal can be located while feeding. Seed extraction involves dental and manual processing of the fruit and is a rather noisy activity. Individual sifaka could easily be located from the ground simply by following the sounds of seeds being cracked and fruit being dropped. Although rufus lemurs do not feed on seeds, they often vocalized while feeding and dropped fruit regularly. In contrast, red-bellied lemurs were difficult to observe while feeding as they moved and fed quietly, rarely dropped fruit while feeding, and, on more than one occasion, a study group was lost because an observer looked away for a moment to write notes.

The fact that sifaka and rufus lemurs exploited similar foods in similar ways suggests that group size may be a more influential factor over body size when selecting resources that are 'safe.' Both exploited ground and lower canopy resources more often, and exploited small and medium patches more often where they were less likely to have a nearest neighbor. Although they were less likely to have a nearest neighbor, other group members fed or rested close by, which may increase detection distance even though they were not feeding together. In contrast, when red-bellied lemurs fed in these kinds of resources, which was not often, they always fed close together.

Conclusions

Overall, sifaka and rufus lemurs fed in riskier locations compared to red-bellied lemurs and the differences between species may be most affected by group size differences. Sifaka and rufus lemurs exploited foods from a wider range of sources: small, medium as well as large patches and ground sources such as herbs and vines. Sifaka and rufus lemurs were not cryptic feeders. Red-bellied lemurs, in contrast, exploited foods primarily from large upper canopy patches and moved and fed quietly.

Acknowledgments

We thank Benjamin Andriamihaja, Patricia Wright, ANGAP (Association Nationale pour la Gestation des Aires Protégées) and the Department of Water and Forests of Madagascar. We are grateful for comments by C. Grassi, L. Miller, A. Treves, M. Rasmussen and two anonymous reviewers regarding improvements to the chapter. We would also like to acknowledge Albert Telo, Pierre Talata, Aimee Razafiarimalala, and the staff of the Ranomafana National Park Office for their assistance. We would also like to acknowledge field assistants: Robert Lessnau, Philip Holman, Matthew Coleflesh, Rachel Kitko, Duncan Fraser, Laura Yandell, Russel Frye, Zoe Forward, and Michele Goodnight. This project was funded by National Science Foundation Grant SBS 9305901, and a National Science Foundation–REU (Research Experience for Undergraduates) supplement.

REFERENCES

Anderson, C.M. (1986). Predation and primate evolution. *Primates* **27**: 15–39.

Baldellou, M., and Henzi, P. (1992). Vigilance, predator detection and the presence of supernumerary males in vervet monkey troops. *Animal Behaviour* **43**: 451–61.

Cheney, D.L., and Wrangham, R.W. (1987). Predation. In: B.B. Smuts, D.L. Cheney, R.W. Seyfarth, R.W. Wrangham, and T.T. Struhsaker, eds., *Primate Societies*. Chicago: University of Chicago Press, pp. 227–39.

Cords, M. (1990). Vigilance and mixed species association of some East African forest monkeys. *Behavioural Ecology and Sociobiology* **26**: 297–300.

Csermely, D. (1996). Antipredator behavior in lemurs: evidence of an extinct eagle on Madagascar or something else? *International Journal of Primatology* **17**: 349–54.

Curtis, D.J., Zaramody, A., and Martin, R.D. (1999). Cathemerality in the mongoose lemur, *Eulemur mongoz*. *American Journal of Primatology* **47**: 279–98.

Dagosto, M. (1995). Seasonal variation in positional behavior of Malagasy lemurs. *International Journal of Primatology* **16**: 807–33.

de Ruiter, J.R. (1986). The influence of group size on predator scanning and foraging behaviour in wedge-capped capuchin monkeys (*Cebus olivaceus*). *Behaviour* **98**: 240–58.

Dollar, L.J., Forward, Z.A., and Wright, P.C. (1997). First study of *Cryptoprocta ferox* in the rainforests of Madagascar. *American Journal of Physical Anthropology* **Suppl. 24**: 102–3.

Eason, P. (1989). Harpy eagle attempts predation on adult howler monkey. *Condor* **91**: 469–70.

Freed, B.Z. (1996). Co-occurrence among crowned lemurs (*Lemur coronatus*) and Sanford's lemurs (*Lemur fulvus sanfordi*) of Madagascar. Ph.D. dissertation, St. Louis, Washington University.

Goodman, S.M. (1994a). The enigma of antipredator behavior in Lemurs: Evidence of a large extinct eagle on Madagascar. *International Journal of Primatology* **15**: 129–34.

Goodman, S.M. (1994b). Description of a new species of subfossil eagle from Madagascar *Stephanoaetus* (Aves: Falconiformes) from the deposits of Ampasambazimba. *Proceedings of the Biological Society, Washington* **107**: 421–8.

Goodman, S.M., O'Connor, S., and Langrand, O. (1993). A review of predation on lemurs. Implications for the evolution of social behavior in small, nocturnal primates. In: P.M. Kappeler and J.U. Ganzhorn, eds., *Lemur Social Systems and Their Ecological Basis*. New York: Plenum Press, pp. 51–63.

Goodman, S.M., Rene de Roland, L.A., and Thorstrom, R. (1998). Predation on the eastern woolly lemur (*Avahi laniger*) and other vertebrates by Henst's goshawk (*Accipiter henstii*). *Lemur News* **3**: 14–15.

Gould, L. (1996). Vigilance behavior during the birth and lactation season in naturally occurring ring-tailed lemurs (*Lemur catta*) at the Beza-Mahafaly Reserve, Madagascar. *International Journal of Primatology* **17**: 331–47.

Gould, L., Fedigan, L.M., and Rose, L.M. (1997). Why be vigilant: The case of the alpha animal. *International Journal of Primatology* **18**: 401–14.

Hemingway, C.A. (1996). Morphology and phenology of seeds and whole fruit eaten by Milne-Edward's sifaka, *Propithecus diadema edwardsi* in Ranomafana National Park, Madagascar. *International Journal of Primatology* **17**: 637–60.

Hemingway, C. A. (1998). Selectivity and variability in the diet of Milne-Edward's sifaka (*Propithecus diadema edwardsi*): Implications for folivory and seed eating. *International Journal of Primatology* **19**: 355–77.

Isbell, L. A. (1994). Predation on primates: Ecological patterns and evolutionary consequences. *Evolutionary Anthropology* **3**: 61–71.

Jolly, A. (1966). *Lemur Behavior*. Chicago: University of Chicago Press.

Macedonia, J.M and Pollock, J.F. (1989). Visual assessment of avian threat in semi-captive ring-tailed lemurs (*Lemur catta*). *Behaviour* **111**: 219–304.

Oda, R. (1998). The response of Verreaux's sifaka to anti-predator alarm calls given by sympatric ring-tailed lemurs. *Folia Primatologica* **69**: 357–60.

Overdorff, D.J. (1993a). Similarities, differences and seasonal patterns in the diets of *Eulemur rubriventer* and *Eulemur fulvus rufus* in the Ranomafana National Park, Madagascar. *International Journal of Primatology* **14**: 721–53.

Overdorff, D.J. (1993b). Reproductive and ecological correlates to ranging patterns in *Eulemur fulvus rufus* and *Eulemur rubriventer* in Madagascar. In: P.M. Kappeler and J.U. Ganzhorn, eds., *Lemur Social Systems and their Ecological Basis*. New York: Plenum Press, pp. 167–78.

Overdorff, D.J. (1996a). Ecological correlates to social structure in two lemur species in Madagascar. *American Journal of Physical Anthropology* **100**: 487–506.

Overdorff, D.J. (1996b). Ecological correlation of activity and habitat use of two Prosimian primates: *Eulemur rubriventer* and *Eulemur fulvus rufus* in Madagascar. *American Journal of Primatology* **40**: 327–42.

Overdorff, D.J., and Strait, S.G. (1995). Life history and predation in *Eulemur rubriventer* in Madagascar. *American Journal of Physical Anthropology* **100**: 487–506.

Overdorff, D.J., and Strait, S.G. (1998). Seed handling by three prosimian primates in southeastern Madagascar: Implications for Seed dispersal. *American Journal of Primatology* **45**: 69–82.

Overdorff, D.J., Merenlender, A.M., Talata, P., Telo, A., and Forward, Z. (1999). Life history of *Eulemur fulvus rufus* from 1988–1997: Implications for ecological stress in southeastern Madagascar. *American Journal of Physical Anthropology* **108**: 295–310.

Peetz, A., Norconk, M.A., and Kinzey, W.G. (1992). Predation by jaguar on howler monkeys (*Alouatta seniculus*) in Venezuela. *American Journal of Primatology* **28**: 223–8.

Rakotondravony, D., Goodman, S.M., and Soarimalala, V. (1998). Predation on *Hapalemur griseus griseus* by *Boa manditra* (Boidea) in the littoral forest of eastern Madagascar. *Folia Primatolologica* **69**: 405–8.

Rasmussen, M.A. (1999). Ecological influences on activity cycle in two cathemeral primates, the mongoose lemur (*Eulemur mongoz*) and the common brown lemur (*Eulemur fulvus fulvus*) Ph.D. dissertation, Durham, Duke University.

Rasoloarison, R.M., Rasolonandrasana, B.P.N., Ganzhorn, J.U., and Goodman, S.M. (1995). Predation of vertebrates in the Kirindy Forest, Western Madagascar. *Ecotropica* **1**: 59–65.

Richard, A. (1978). *Behavioral Variation*. New York: Freeman Press.

Rose L.M., and Fedigan, L.M. (1995). Vigilance in white-faced capuchins, *Cebus capuchinus* in Costa Rica. *Animal Behaviour* **49**: 63–70.

Sauther, M.L. (1989). Antipredator behaviour in troops of free-ranging

Lemur catta at Beza Mahafaly Special Reserve, Madagascar. *International Journal of Primatology* **10**: 595–606.

Schwab, D. (1999). Predation on *Eulemur fulvus* by *Accipiter henstii* (Henst's goshawk). *Lemur News* **4**: 34.

Skorupa, J. (1989). Crowned eagles *Stephanoaetus coronatus* in rainforest: observations on breeding chronology and diet at a nest in Uganda. *Ibis* **121**: 294–8.

Strait, S.G., and Overdorff, D.J. (1995). Fracture toughness of plants eaten by *Propithecus diadema edwardsi*. *American Journal of Physical Anthropology* **Suppl. 20**: 206.

Struhsaker, T.T., and Leakey, M. (1990). Prey selectivity by crowned hawk-eagles on monkeys in the Kibale Forest, Uganda. *Behavioural Ecology and Sociobiology* **26**: 435–43.

Treves, A. (1998). The influence of group size and neighbors on vigilance in two species of arboreal monkeys. *Behaviour* **135**: 453–81.

van Schaik, C.P., and Kappeler, P.M. (1996). Social systems of gregarious lemurs: lack of convergence with anthropoids due to evolutionary disequilibrium? *Ethology* **102**: 915–41.

van Schaik, C. P., van Noordwijk, M.A., Warsono, B., and Surtiono, E. (1983). Party size and early detection of predators in Sumatran forest primates. *Primates* **24**: 211–21.

Wilson, J. M., Stewart, P. D., Ramangason, G.-S., Denning, A. M., and Hutchings, M. S. (1989). Ecology and conservation of the crowned lemur, *Lemur coronatus*, at Ankarana, N. Madagascar. *Folia Primatologica* **52**: 1–26.

Wright, P.C. (1992). Primate ecology, rainforest conservation, and economic development: Building a national park in Madagascar. *Evolutionary Anthropology*, **1**: 25–33.

Wright, P.C. (1995). Demography and life-history of free-ranging *Propithecus diadema edwardsi* in Ranomafana National Park, Madagascar. *International Journal of Primatology* **16**: 835–54.

Wright, P.C. (1998). Impact of predation risk on the behaviour of *Propithecus diadema edwardsi* in the rain forest of Madagascar. *Behaviour* **135**: 483–512.

Wright, P.C., and Martin, L.B. (1995). Predation, pollination and torpor in two nocturnal primates: *Cheirogaleus major* and *Microcebus rufus* in the rain forest of Madagascar. In: L. Alterman, G. Doyle, and M. K. Izard, eds., *Creatures of the Dark*. New York: Plenum Press, pp. 45–60.

Wright, P.C., Heckscher, S.K., and Dunham, A.E. (1997). Predation on Milne-Edward's sifaka (*Propithecus diadema edwardsi*) by the fossa (*Cryptoprocta fossa*) in the rainforest of Southeastern Madagascar. *Folia Primatologica* **68**: 34–43.

Yamashita, N. (1996). Seasonality and site specificity of mechanical dietary patterns in two Malagasy lemur families (Lemuridae and Indriidae). *International Journal of Primatology* **17**: 355–87.

9 • Evidence of predator sensitive foraging and traveling in single- and mixed-species tamarin troops

PAUL A. GARBER & JÚLIO CÉSAR BICCA-MARQUES

Introduction

The threat of predation and behavioral tactics associated with predator avoidance are reported to play a critical role in primate socioecology (Hill and Dunbar 1998, Terborgh and Janson 1986, van Schaik 1983). In particular, it has been suggested that predation risk can have a significant effect on group size and composition (Cheney and Wrangham 1987, Isbell 1994, Stanford 1998, Treves 1999), vigilance behavior (Burger and Cochfeld 1994), patterns of habitat utilization (Cowlishaw 1999), within-group spacing, and individual foraging success (Cowlishaw 1998). Hill and Dunbar (1998: 412) define predation risk as 'the animals' own perception of the likelihood of being subject to an attack by a predator . . . it reflects the animals' collective past historical experience of actual attacks by predators and is the basis on which the animals implement their antipredator strategies.' In social animals, individuals may assess predation risk based on their own personal information, as well as by relying on alarm calls and vigilance behavior provided by other group members. Several authors have suggested that when relying on group-based information, including shared vigilance, individuals living in larger social groups experience lower predation risk than conspecifics living in smaller social groups (Janson and Goldsmith 1995, Terborgh 1990, Terborgh and Janson 1986, van Schaik 1983). The mechanisms promoting shared or cooperative vigilance, however, are poorly understood (Lima and Bednekoff 1999).

Tamarin monkeys

In this chapter, we examine evidence of predator sensitive foraging in wild tamarins in an attempt to link social interactions, foraging patterns, and predator avoidance behaviors. Specifically, we examine evidence of antipredator behavior in single and mixed-species troops of saddleback (*Saguinus fuscicollis weddelli*) and emperor (*S. imperator imperator*) tamarins when foraging at experimental feeding platforms. We refer to this as foraging in 'small-scale space.' We also examined patterns of ranging and travel routes taken to reach sequential feeding trees in mixed-species troops of saddleback (*S. fuscicollis nigrifrons*) and moustached (*S. mystax mystax*) tamarins. We refer to this as travel in 'large-scale space.' Predation risk and antipredator tactics used by tamarins when foraging within a food patch or tree crown may differ considerably from predation risk and antipredator tactics used when traveling between distant feeding sites.

Tamarins (genus *Saguinus*) represent a highly successful radiation of small-bodied (300–550 g) New World monkeys that are characterized by high levels of social cooperation in infant care, especially among adult male group members (Caine 1993, Garber 1988, 1997, Garber *et al.* 1993, Goldizen 1990). All species are reported to live in cohesive groups of approximately five to ten individuals (Garber 1994, 1997, Snowdon and Soini 1988). Due to their small group size, diminutive body size, and relatively long daily path length, tamarins are potentially vulnerable to a wide range of predators (Caine 1993, Goldizen 1987, Heymann 1990, Moynihan 1970, Peres 1993, Terborgh 1983) including snakes, felids, procyonids, and, especially, raptors. Although sightings of successful predatory attacks on tamarins are rare (Table 9.1), Caine (1993: 201) has argued 'that predation is among the most important, if not the most important, selection pressure influencing the social behavior and group structure of tamarin species.'

In addition, several species of tamarins (*Saguinus fuscicollis, S. labiatus, S. imperator, S. mystax*, and, possibly, *S. nigricollis*) form mixed-species troops. These associations are extremely stable, with individuals of each species feeding, foraging, resting, and traveling together throughout the year (Garber 1988, Izawa 1978, Peres 1993, Snowdon and Soini 1988, Terborgh 1983). Researchers have argued that, because tamarin mixed-species troops contain more individuals than tamarin single-species groups, this association provides fitness benefits through increased predator vigilance and predator detection (Hardie and Buchanan-Smith 1997, Heymann 1990, Norconk 1986, Peres 1993, Ramirez 1989). However, we question the degree to which tamarin mixed-species troops provide a significant antipredator advantage to its members. Specifically, we found that

Table 9.1. *Reported cases of predation, predator attacks, and the occurrence and frequency of alarm call events by tamarins when in mixed-species troops and single-species groups*

Species	Length of study[a]	Observed predation events	Predator attacks	Alarm call events	References
Mixed-species troops					
S. imperator	10+ yrs	1	1/1–2 weeks	>1 alarm hr^{-1}	1, 2
S. fuscicollis	10+ yrs	1			
S. imperator	19 mo.	0	several per week		3
S. fuscicollis	19 mo.	0	several per week		
S. imperator	6 mo. (140 h)	0	0		19
S. fuscicollis	6 mo. (175 h)	0	0		
S. mystax	14 mo. (731 h)	0	1 per 8.8 days		4
S. fuscicollis	14 mo. (731 h)	0			
S. mystax	11 mo. (417 h)	0	1 (raptor)	0.5 alarm h^{-1}	5
S. fuscicollis	11 mo. (417 h)	0		0.3 alarm h^{-1}	
S. mystax	12 mo.	0		1 alarm response to terrestrial carnivores; numerous alarm responses to avians	6
S. fuscicollis	12 mo.	0			
S. mystax	17 mo.	0	0		7
S. fuscicollis	17 mo.	0	0		
S. mystax	13 mo.	0	3 (coati to live tamarin in cage; tayra sighted nearby; raptor attack)	several alarm calls per day (1 case of mobbing boa)	8
S. fuscicollis	13 mo.	0			
S. mystax	11 mo. (627 h)	0	3 (raptor)	several alarm call events per day at raptors, toucans, other birds, tayra (3), squirrel, and deer	9
S. fuscicollis	11 mo. (627 h)	0	1 (toucan)	1 alarm event to an ocelot	
S. labiatus	5 mo.	0	2 (mob tayra)	alarm calls at raptors and other birds	21
S. fuscicollis	5 mo.	0			

Species		Predation	Predation	Response	Ref.
S. labiatus	4 mo.	0	1 (raptor)	alarm calls at raptors	10
S. fuscicollis	4 mo.	0			
Callimico goeldii	4 mo.	0			
Single-species groups					
S. mystax	13 mo. (433 h)	0	0	0.3 alarm h^{-1}	5
S. mystax	3+ mo.	1 (anaconda)	0		11
S. mystax	4 mo.	0	0		9
S. geoffroyi	9 mo.	0	0	1 mobbing of coatimundi	12
S. geoffroyi	16 mo.	0	1 (toucan)	alarm responses to avians	13
S. oedipus	24 mo.	0	1 (raptor)	7 alarm responses to tayra in area	14
S. oedipus	9 mo. (403 h)	0	0	alarm call to large birds	15
S. nigricollis	7 mo.	1 (a second predation event was observed during an earlier study)	18 (raptor) (these could be classified as alarm call events)	alarm at raptors and terrestrial carnivores	16
S. nigricollis	9 mo.	0	0	alarm response to raptors	20
S. fuscicollis	5 mo.	0	1+ (raptor)	alarm response to raptors and terrestrial mammals	17
S. fuscicollis	15 mo. (416 h)	0	0	alarm reactions to tayra and raptors	18
Total	582 mo.	4[b]			

Notes:

[a] Figures in parentheses are the hours of observation in a study.

[b] A fifth successful predation event was observed but not during the study.

Source: 1, Terborgh 1983; 2, Goldizen 1987; 3, Windfelder 1997; 4, Peres 1991; 5, Heymann 1990; 6, Garber unpubl. data; 7, Castro 1991; 8, Ramirez 1989; 9, Norconk 1986 and unpubl. data; 10, Garber unpubl. data; 11, Bartecki and Heymann 1987; 12, Garber unpubl. data; 13, Dawson 1976; 14, Neyman 1978; 15, Savage 1990; 16, Izawa 1978; 17, Crandlemire-Sacco 1986; 18, Soini 1987; 19, Bicca-Marques unpubl. data; 20, de la Torre et al. 1995; 21, Buchanan-Smith 1990.

tamarins in single-species groups did not forage in a more predator sensitive way than did the same individuals when part of a
larger mixed-species troop. When traveling in large-scale space,
mixed-species tamarin troops employed a set of antipredator
tactics associated with crypticity. However, there was no evidence of
shared or cooperative vigilance between species. Rather, in each
species, a single individual, often a dominant adult male, acted as
a sentinel scanning the environment, presumably for predators.
We present our data on predator sensitive foraging and traveling
below.

Methods

Data on tamarin feeding, ranging, and vigilance behavior were collected on single- and mixed-species troops of emperor and saddleback tamarins in Brazil (Parque Zoobotanico, Federal University of
Acre, Rio Branco, 9° 56′S, 67° 53′W), and mixed-species troops of
moustached and saddleback tamarins in northeastern Peru (Rio
Blanco, 4° 15′S, 73° 04′W). Tamarins at both sites were trapped and
marked with beaded identification collars.

Research in Brazil was conducted by JCBM and represents an
experimental field study designed to examine the ability of wild
tamarins to use social and ecological information in within-patch
foraging decisions. The protocol for this study involved the construction of four feeding stations each composed of eight feeding
platforms. Two of eight platforms at each feeding station were
baited with real bananas. The remaining six platforms were baited
with either plastic bananas or real bananas inside a wire mesh cage
(sham reward). Platforms were baited for a period of 125 consecutive days (see Bicca-Marques 2000). Data presented represent 9671
visits by individually marked tamarins at our feeding platforms.
The saddleback tamarin study group (hereafter referred to as FUS
'A') contained five animals including one adult female and two
adult males. The emperor tamarin study group (hereafter referred
to as IMP 'B') also contained five animals including two adult
females and two adult males.

A group was considered to visit a feeding station as a single-
species group when (a) it arrived out-of-association at a feeding
station that had not been visited by other groups since the last
baiting, or (b) the associating species did not arrive until all feeding
platforms had been visited. A visit was scored as a mixed-species
troop when (a) a group inspected at least one feeding platform

while its associating group was also visiting a feeding platform, or
(b) if the group had not yet visited all platforms when the associating group arrived at the feeding station.

Data in Peru were collected by PAG and represent 34 complete
days of observations (17 days for moustached tamarins and 17 days
for saddleback tamarins). The saddleback tamarin group contained
seven animals, including three adult males and two adult females.
The moustached tamarin group contained eight animals, including two adult males and two adult females.

The tamarin study troop was followed from the time it left its
sleeping tree in the morning until the time it entered its sleeping
tree at night. Thus, our unit of study was a complete tamarin day,
and for each day a group activity budget, dietary profile, and
pattern of ranging was constructed for each species of the troop.
Information reported here is based on 4396 individual activity
records for moustached tamarins and 4351 individual activity
records for saddleback tamarins (Garber 1988, 1993, 1998).

Results

Foraging in small-scale space

If predation risk has a significant effect on within-patch foraging
behavior, we expect that when arriving at a feeding station as a
single-species group, tamarins would be more cautious and more
vigilant than when arriving at the same feeding station as part of a
larger mixed-species troop. This would support the hypothesis
that, through cooperative vigilance, individuals in larger groups
would benefit by increased predator detection and reduced perceived levels of predation risk at feeding sites. This is especially true
given that our feeding platforms were in relatively open areas of
the forest and that the tamarins visited these feeding stations
several times each day.

After arriving at a feeding station, both emperor and saddleback
tamarins quickly descended to the platforms and began to feed
(within 1–2 minutes of arrival, Table 9.2). Using latency as a
measure of perceived predation risk, there was little evidence that
group size had a significant effect on tamarin foraging behavior. In
FUS 'A' and IMP 'B' latency of group arrival at a feeding station to
time of first feeding event did not differ significantly ($p>0.32$ for
FUS 'A' and $p>0.75$ for IMP 'B') when arriving as a single-species
group or as part of a mixed-species troop.

If perceived predation risk was higher when foraging as a smaller

Table 9.2. *Comparison of tamarin foraging behavior when visiting feeding stations as a single-species group and as a mixed-species troop. Mean values shown, sample sizes in parentheses*

Study group association state	IMP 'B'		FUS 'A'	
	Single	Mixed	Single	Mixed
Latency of group arrival at feeding station to first feeding event (min)	1.1 (212)	1.7 (35)	1.5 (266)	2.2 (74)
Latency of last feeding event to departure from the feeding station (min)	2.4 (212)	2.4 (31)	6.4 (266)	5.6 (73)
Total time spent at feeding station (min)	10.4 (221)	9.7 (41)	16.1 (270)	15.9 (84)
Time spent on feeding platforms (min)	[a]6.0 (221)	5.1 (26)	[a]6.3 (270)	3.7 (73)
Number of group members present at feeding station	4.3 (221)	4.3 (41)	[a]4.1 (270)	3.7 (84)
Number of group members visiting a feeding platform	3.9 (220)	3.9 (41)	[a]3.9 (269)	2.8 (84)
Maximum number of group members feeding together on a platform	2.2 (221)	2.2 (20)	[a]3.4 (210)	2.6 (50)

Notes:
[a] $p < 0.05$.

single-species group, then tamarins are expected to spend less time at the feeding station, less time on feeding platforms, and leave the feeding station more quickly after the last feeding event than when foraging as part of a mixed-species troop. Our results indicate, however, that IMP 'B' spent significantly more time at the feeding station, more time feeding on platforms, and remained longer at the feeding stations after the last animal fed when arriving as a single-species group than when in association with saddleback tamarins (Table 9.2). A similar behavioral pattern was observed in FUS 'A', although in this group the difference between results in-association and out-of-association did not reach statistical significance (Table 9.2).

It is possible, however, that the number of individuals feeding together on a platform is a more sensitive measure of perceived predation risk than the total time spent by all group members on a feeding platform. This results from the fact that, as the number of foragers decreases, the number of group members potentially available to monitor and scan the environment for predators increases. In emperor tamarin group IMP 'B', however, there were no significant differences in the number of group members visiting feeding platforms, or in the maximum number of group members feeding together on experimental platforms when foraging in mixed-species troops compared to single-species groups (Table 9.2). In FUS 'A', each of these variables was significantly different between single- and mixed-species conditions, but not in a direction consistent with an antipredator vigilance strategy. When FUS 'A' arrived alone, there were more individuals at the feeding station, more individuals on feeding platforms, more individuals feeding together, and, therefore, fewer animals to be vigilant than when FUS 'A' arrived as part of a mixed-species troop. Overall, there was no evidence that either emperor tamarins or saddleback tamarins showed greater caution or increased vigilance when foraging as a single-species group.

Sentinel behavior

In emperor and saddleback tamarins, a group member acted as a sentinel while at the feeding station. This individual scanned the environment and waited for others to feed before descending to the platforms. In IMP 'B', the most dominant group member, adult male IB-VRM (based on directionality of aggressive interactions, Bicca-Marques 2000) was generally the last individual to descend to feed (Table 9.3). In FUS 'A', adult male AZL, along with adult female ROS, were the highest ranking group members. FA-AZL was frequently the last adult group member to arrive on the feeding platforms (Table 9.3). Although FA-AZL's latency to visit platforms was influenced by the fact that he was observed to carry infant(s) more often than other group members (Bicca-Marques unpublished data), he also exhibited more intense vigilance activities. Sentinel behavior by dominant male tamarins may be best understood as part of a suite of behaviors associated with infant care and predator vigilance.

Predator sensitive traveling in large-scale space

Data on travel and ranging in mixed-species troops of moustached and saddleback tamarins in northeastern Peru were examined to identify evidence of predation risk and antipredator behavior. Travel accounted for approximately 30% of the tamarin activity budget. In general, the tamarins traveled in single-line progression, with most or all group members using the same set of travel routes and arboreal pathways. Movement between feeding sites appeared to be goal-directed. Troop members commonly took relatively straight-line and direct travel routes to reach distant feeding sites (Garber 1989). Each group traveled an average of 1400 m per day,

Table 9.3. *A comparison of individual patterns of foraging behavior in emperor tamarin group IMP 'B' and saddleback tamarin group FUS 'A'*

Foraging behavior	Individual (IMP 'B')				
	VRM	VEP	REP	AZL	PRO
Latency of first visit to a feeding platform after group arrival (min)	[a]3.3±2.8 (N=249)	2.3±2.3 (N=273)	1.7±2.7 (N=278)	2.4±3.1 (N=309)	1.6±2.8 (N=269)
Time spent feeding at reward platforms (min)	2.2±1.3 (N=244)	1.6±1.0 (N=267)	2.2±1.3 (N=332)	1.4±1.0 (N=307)	2.0±1.1 (N=332)
Percentage of total reward visits when she/he found the reward platform	9.9 (N=46)	22.1 (N=103)	13.1 (N=61)	24.7 (N=115)	30.3 (N=141)

	Individual (FUS 'A')				
	AZL	BRA	ROS	AMA	INF1
Latency of first visit to a feeding platform after group arrival (min)	[a]3.3±5.6 (N=406)	1.6±3.1 (N=454)	2.7±4.8 (N=416)	2.4±3.0 (N=441)	4.5±7.7 (N=123)
Time spent feeding at reward platforms (min)	2.5±1.5 (N=435)	2.6±1.5 (N=548)	2.1±1.2 (N=472)	2.4±1.3 (N=523)	2.4±1.6 (N=127)
Percentage of total reward visits when she/he found the reward platform	20.9 (N=123)	46.1 (N=271)	27.5 (N=162)	5.2 (N=31)	NA

Notes:
Values shown are means ± standard deviations, and sample sizes (in parentheses). [a] $p < 0.05$. NA = not applicable.

and visited between 12 and 13 individual feeding trees (Table 9.4). Mean travel speed was 6–12 m per minute (Table 9.4).

During travel, the attention of group members was directed to their immediate surroundings (following the lead individual, identifying branches of appropriate size and weight bearing capacity for continued travel, and navigating gaps in the canopy), and possibly to landmarks off in the distance (spatial orientation). Given the denseness of the rainforest canopy, it seems unlikely that individuals were able to monitor the vigilance activities of other group members during high-speed progression, or to detect predators as effectively as when engaged in more sedentary activities. If this assumption is correct, then what types of behaviors do tamarins use to increase their ability to detect and avoid predators during travel?

Table 9.4. *Patterns of ranging and habitat utilization in a mixed-species tamarin troop*

Species	Day range (m)	Number of feeding trees visited	Travel bout length (min)	Rate of travel (m min^{-1})	Travel bout distance (m)	Percentage of travel bouts ≤2 min	Distance traveled between feeding trees (m)
S. mystax	1427	12.4	4.2	5.6–9.0	23.5–37.8	47.7	88.0. ±14.3
S. fuscicollis	1380	13.5	3.8	5.7–11.7	21.6–42.5	50.8	81.9. ±17.0

We examined this question by analyzing the duration and distance of travel bouts. A travel bout represented a consecutive set of activity records in which the individual was engaged in travel. If travel between feeding sites is characterized by long bouts of uninterrupted movement, then the ability of individuals to effectively scan the environment for predators is limited. In contrast, if travel bouts tend to be short and marked by brief periods of inactivity, individuals have an increased opportunity to use behavioral crypticity and vigilance in predator detection. As indicated in Table 9.4, both tamarin species were characterized by brief sequences of travel. In moustached tamarins, over 47% (309/647) of all travel bouts were ≤2 minutes in duration. In saddleback tamarins, 51% (269/529) of travel bouts were ≤2 minutes in duration. Based on their rate of arboreal progression, moustached and saddleback tamarins traveled a distance of 22–42 m before either stopping to rest, or changing to a vigilance-compatible activity such as grooming or resting (Table 9.4). Given that the average distance between sequential feeding sites was 85 m (±15.6 m), tamarins temporarily halted travel activities two to four times on route to their next feeding tree.

We also found that members of our mixed-species study troop rarely reused the same locomotor pathways or parts of their home range on consecutive days, and utilized a large number of different travel routes to reach frequently revisited feeding trees. This reduced the predictability of group movements in time and space. Overall, when moving between distant feeding sites, both moustached and saddleback tamarins employed a pattern of behavioral crypticity that was consistent with predator sensitive travel.

Discussion

Several hypotheses have been offered to account for the potential effects of predation on group size, group composition, and social behavior in primates (Hill and Dunbar 1998). It is generally assumed that individuals living in larger groups receive higher fitness benefits in terms of reduced vulnerability to predators than do individuals living in smaller groups (Terborgh and Janson 1986). Larger group size offers more eyes and ears for predator detection, and through a pattern of coordinated or shared vigilance may reduce the time and energy spent per individual in antipredator behaviors. According to a 'shared vigilance' model, as the total number of vigilant neighbors increases, nonvigilant individuals can devote more time to foraging and other maintenance activities (Pulliam, 1973). This presupposes a direct relationship between the number of animals vigilant and a group's ability to detect predators. However, studies by Cords (1990) on arboreal guenons (*Cercopithecus mitis* and *C. ascanius*), Rose and Fedigan (1995) on white-faced capuchins (*Cebus capucinus*), Cowlishaw (1998) on desert baboons (*Papio cynocephalus ursinus*), and Treves (1998) on redtail monkeys (*Cercopithecus ascanius*) and red colobus monkeys (*Procolobus badius*) report no significant effect of group size on vigilance behavior. Moreover, Treves (1998: 471) found that in *C. ascanius* and *P. badius* 'individuals in groups associated with other species did not decrease time spent scanning when compared to the same groups in single-species state.' Also, although Isbell and Young (1993) report that rates of scanning decreased with increasing group size in vervets (*Cercopithecus aethiops*), group size had no effect on rates of predation. These findings are consistent with Hill and Dunbar's (1998) recent analyses in which they report no evidence of a significant relationship between group size and predation rate in a sample of 19 species of diurnal primates.

Despite the fact that mixed-species tamarin troops contained nearly twice the number of animals as single-species groups, troop size had no discernible effect on predator sensitive foraging behavior. In general, tamarins spent more time at a feeding station, more time on feeding platforms, and had more animals feeding simultaneously when in single-species groups than did these same individuals when part of a larger mixed-species association. During within-patch foraging, tamarin single-species groups were neither more cryptic nor more vigilant than were mixed-species troops. It appears that individual tamarins rely primarily on the sentinel behavior of a single group member, along with each individual's own perception of risk and vigilance behavior, to detect predators when at a feeding site. There was no evidence of shared vigilance.

In large-scale space, however, tamarins may face more severe problems of predator detection. This is because high-speed travel and continuous progression are incompatible with effective forms of vigilance. When moving between distant feeding sites, the mixed-species tamarin troop was characterized by relatively short sequences of travel followed by brief pauses in which animals scanned the environment, probably for potential predators, landmarks, travel routes, resources, and the presence of neighboring groups. Given that tamarins are vulnerable to a wide range of terrestrial, arboreal, and aerial predators, we argue that tamarin movements in large-scale space (e.g., short travel bouts, limited reuse of the same arboreal pathways, avoiding areas of their range already visited that same day) are consistent with a pattern of predator avoidance, and behavioral crypticity.

A final issue we address is the likelihood that predation risk has played a primary role in shaping tamarin socioecology. As indicated in Table 9.1, at study sites such as Manu National Park, Peru (Terborgh 1983, Goldizen 1987, Windfelder 1997) and Urucu River, Brazil (Peres 1991, 1993) predation risk for tamarins appears to be extremely high. At these sites, direct predator attacks on tamarins are reported to occur from one to several times per week, despite the fact that the tamarins form mixed-species troops. At other sites where mixed-species troops of tamarins also have been studied (Rio Blanco, Peru), observed predator attacks average less than one per year (Table 9.1). These latter values are comparable with data on predation in most single-species tamarin groups. Given such variance in predation risk between study sites, as well as among tamarins residing in single- and mixed-species troops, it remains uncertain whether tamarin mixed-species troops evolved primarily in response to predation pressure.

Acknowledgments

We wish to thank Lynne Miller for inviting us to contribute to this volume and for the insightful comments she, her dad, and the anonymous reviewers provided on earlier drafts of this manuscript. PAG wishes to thank Eriberto Mermao, Walter Mermao, and Ursula Iwaniec for their assistance in the field. In addition, Drs Luis Moya, Jaimie Moro, Filomeno Encarnación, and Carlos Malaga provided logistical support in the field. Funds to support this research were provided through grants from the National Science Foundation (BNS-8310480) and the Research Board of the University of Illinois. Dr Marilyn Norconk kindly allowed us to cite her unpublished data on predation rates in *Saguinus*. PAG wishes to acknowledge Sara Garber's and Jenni Garber's valuable insights on predator sensitive travel in large-scale space, small-scale space, and when crossing the street.

JCBM thanks his wife Cláudia and his son Gabriel for all their support and encouragement, his research assistants (Karin, Claudio, and Alice) for helping in the data collection, and the people from the Universidade Federal do Acre and Fundação S.O.S. Amazonia for their logistical support. He also thanks the Brazilian Higher Education Authority (CAPES) for the doctoral fellowship that supported his studies at the University of Illinois and the financial support provided by the World Wildlife Fund – Brazil, Fundação O Boticário de Proteção à Natureza, Wenner-Gren Foundation for Anthropological Research, American Society of Primatologists, and Tinker Fund/Center for Latin American and Caribbean Studies-UIUC.

REFERENCES

Bartecki, U., and Heymann, E.W. (1987). Field observation of snake-mobbing in a group of saddleback tamarins, *Saguinus fuscicollis nigrifrons*. *Folia Primatologica* **48**: 199–202.

Bicca-Marques, J.C. (2000). Cognitive aspects of within-patch foraging decisions in wild diurnal and nocturnal New World monkeys, Ph.D. thesis, University of Illinois, Urbana.

Buchanan-Smith, H. (1990). Polyspecific association of two tamarin species, *Saguinus labiatus* and *Saguinus fuscicollis*, in Bolivia. *American Journal of Primatology* **22**: 205–14.

Burger, J., and Cochfeld, M. (1994). Vigilance in African mammals: differences among mothers, other females, and males. *Behaviour* **131**: 153–69.

Caine, N.G. (1993). Flexibility and co-operation as unifying themes in *Saguinus* social organisation and behaviour: The role of predation pressure. In: A.B. Rylands, ed., *Marmosets and Tamarins: Systematics, Behaviour, and Ecology*, Oxford: University Press, Oxford, pp. 200–19

Castro, R. (1991). Behavioral ecology of two coexisting tamarin species (*Saguinus fuscicollis nigrifrons* and *Saguinus mystax mystax*, Callitrichidae, Primates) in Amazonian Peru, Ph.D. thesis, Washington University, St. Louis.

Cheney, D.L., and Wrangham, R.W. (1987). Predation. In: B.B. Smuts, D.L. Cheney, R.M. Seyfarth, R.W. Wrangham, and T.T. Struhsaker, eds., *Primate Societies*. Chicago: University of Chicago Press, pp. 227–39.

Cords, M. (1990). Vigilance and mixed-species associations of some East African forest monkeys. Behavioral Ecology and Sociobiololgy **26**: 297–300.

Cowlishaw, G. (1998). The role of vigilance on the survival and reproductive strategies of desert baboons. *Behaviour* **135**: 431–52.

Cowlishaw, G. (1999). Trade-offs between foraging and predation risk determine habitat use in a desert baboon population. *Animal Behavior* **53**: 667–86.

Crandlemire-Sacco, J.L. (1986). The Ecology of the Saddle-Backed Tamarin, *Saguinus fuscicollis*, of Southeastern Peru, Ph.D. thesis, University of Pittsburgh.

Dawson, G.A. (1976). *Behavioral ecology of the Panamanian Tamarin, Saguinus oedipus*, Ph.D. thesis, Michigan State University, East Lansing.

de la Torre, S., Campos, F., and de Vries, T. (1995). Home range and birth seasonality of *Saguinus nigricollis graellsi* in Ecuadorian Amazonia. *American Journal of Primatology* **37**: 39–56.

Garber, P.A. (1988). Diet, foraging patterns, and resource defense in a mixed species troop of *Saguinus mystax* and *Saguinus fuscicollis* in Amazonian Peru. *Behaviour* **105**: 18–33.

Garber, P.A. (1989). The role of spatial memory in primate foraging patterns: *Saguinus mystax* and *Saguinus fuscicollis*. *American Journal of Primatology* **19**: 203–16.

Garber, P.A. (1993). Seasonal patterns of diet and ranging in two species of tamarin monkeys: stability versus variability. *International Journal of Primatology* **14**: 145–66.

Garber, P.A. (1994). Phylogenetic approach to the study of tamarin and marmoset social systems. *American Journal of Primatology* **34**: 199–219.

Garber, P.A. (1997). One for all and breeding for one: cooperation and competition as a tamarin reproductive strategy. *Evolutionary Anthropology* **5**: 187–99.

Garber, P.A. (1998). Within- and between-site variability in moustached tamarin (*Saguinus mystax*) positional behavior during food procurement. In: E. Strasser, J. Fleagle, A. Rosenberger, and H. McHenry, eds., *Primate Locomotion: Recent Advances*. New York: Plenum Press, pp. 61–78.

Garber, P.A., Pruetz, J.D., and Isaacson, J. (1993). Patterns of range use, range defense, and intergroup spacing in moustached tamarin monkeys (*Saguinus mystax*). *Primates* **34**: 11–25.

Goldizen, A.W. (1987). Tamarins and marmosets: communal care of offspring. In: B.B. Smuts, D.L. Cheney, R.M. Seyfarth, R.W. Wrangham, and T.T. Struhsaker, eds., *Primate Societies*. London:University of Chicago Press, pp. 34–43.

Goldizen, A.W. (1990). A comparative perspective on the evolution of tamarin and marmoset social systems. *International Journal of Primatology* **11**: 63–83.

Hardie, S.M., and Buchanan-Smith, H.M. (1997). Vigilance in single- and mixed-species groups of tamarins (*Saguinus labiatus* and *Saguinus fuscicollis*). *International Journal of Primatology* **18**: 217–34.

Heymann, E.W. (1990). Reactions of wild tamarins, *Saguinus mystax* and

Saguinus fuscicollis, to avian predators. *International Journal of Primatology* **11**: 327–38.

Hill, R.A., and Dunbar, R.I.M. (1998). An evaluation of the roles of predation rate and predation risk as selective pressures on primate grouping behaviour. *Behaviour* **135**: 411–30.

Isbell, L.A. (1994). Predation on primates: ecological patterns and evolutionary consequences. *Evolutionary Anthropology* **3**: 61–71.

Isbell, L.A., and Young, T.P. (1993). Social and ecological influences on activity budgets of vervet monkeys, and their implications for group living. *Behavioral Ecology and Sociobiology* **32**: 377–85.

Izawa, K. (1978). A field study of the ecology and behavior of the black-mantle tamarin (*Saguinus nigricollis*). *Primates* **19**: 241–74.

Janson, C.H., and Goldsmith, M.L. (1995). Predicting group size in primates: foraging costs and predation risks. *Behavioral Ecology* **6**: 326–36.

Lima, S.L., and Bednekoff, P.A. (1999). Back to the basics of antipredatory vigilance: can nonvigilant animals detect attack? *Animal Behavior* **58**: 537–43.

Moynihan, M. (1970). Some behavior patterns of platyrrhine monkeys II. *Saguinus geoffroyi* and some other tamarins. *Smithsonian Contributions to Zoology* **28**: 1–77.

Neyman, P.F. (1978). Aspects of the ecology and social organization of free-ranging cotton-top tamarins (*Saguinus oedipus*) and the conservation status of the species. In: D.G. Kleiman, ed., *The Biology and Conservation of the Callitrichidae*, Washington, DC: Smithsonian Institution Press, pp. 39–72.

Norconk, M.A. (1986). Interactions between primate species in a neotropical forest: mixed-species troops of *Saguinus mystax* and *S. fuscicollis* (Callitrichidae), Ph.D. thesis, University of California, Los Angeles.

Peres, C.A. (1991). Ecology of mixed-species groups of tamarins in Amazonian Terra Firme forests, Ph.D. thesis, University of Cambridge, Cambridge.

Peres, C.A. (1993). Anti-predation benefits in a mixed-species group of Amazonian tamarins. *Folia Primatologica* **61**: 61–76.

Poucet, B. (1993). Spatial cognitive maps in animals: new hypotheses on their structure and neural mechanisms. *Psychological Review* **100**: 163–82.

Pulliam, H.R. (1973). On the advantages of flocking. *Journal of Theoretical Biology* **38**: 419–22.

Ramirez, M.M. (1989). Feeding ecology and demography of the moustached tamarin *Saguinus mystax* in Northeastern Peru, Ph.D. thesis, City University of New York, New York.

Rose, L.M., and Fedigan, L.M. (1995). Vigilance in white-faced capuchins, *Cebus capucinus*, in Costa Rica. *Animal Behavior* **49**: 63–70.

Savage, A. (1990). The reproductive biology of the cotton-top tamarin

(*Saguinus oedipus oedipus*) in Colombia, Ph.D. Thesis, University of Wisconsin, Madison.

Savage, A., Snowdon, C.T., Giraldo, L.H., and Soto, L.H. (1996). Parental care patterns and vigilance in wild cotton-top tamarins (*Saguinus oedipus*). In: M.A. Norconk, A.L. Rosenberger, and P.A. Garber, eds., *Adaptive Radiations of Neotropical Primates*. New York: Plenum Press, pp. 187–99.

Snowdon, C.T., and Soini, P. (1988). The tamarins, genus *Saguinus*. In: R.A. Mittermeier, A.B. Rylands, A.F. Coimbra-Filho and G.A.B. Fonseca, eds., *Ecology and Behavior of Neotropical Primates*, Vol. 2. Washington, DC: World Wildlife Fund, pp. 223–98.

Soini, P. (1987). Ecology of the saddleback tamarin *Saguinus fuscicollis illigeri* on the Rio Pacaya, Northeastern Peru. *Folia Primatologica* **49**: 11–32.

Stanford, C.B. (1998). Predation and male bonds in primate societies. *Behaviour* **135**: 513–33.

Terborgh, J. (1983). *Five New World Primates*, Princeton NJ: Princeton University Press.

Terborgh, J. (1990). Mixed flocks and polyspecific associations: costs and benefits of mixed groups to birds and monkeys. *American Journal of Primatology* **21**: 87–100.

Terborgh, J., and Janson, C.H. (1986). The socioecology of primate groups. *Annual Review of Ecology and Systematics* **17**: 111–35.

Treves, A. (1998). The influence of group size and neighbors on vigilance in two species of arboreal monkeys. *Behaviour* **135**: 453–81.

Treves, A. (1999). Has predation shaped the social system of arboreal primates? *International Journal of Primatology* **20**: 35–67.

van Schaik, C.P. (1983). Why are diurnal primates living in groups? *Behaviour* **87**: 120–44.

Windfelder, T.L. (1997). Polyspecific association and interspecific communication between two neotropical primates: saddleback tamarins (*Saguinus fuscicollis*) and Emperor Tamarins (*Saguinus imperator*), Ph.D. Thesis, Duke University, Durham.

10 • Predator (in)sensitive foraging in sympatric female vervets (*Cercopithecus aethiops*) and patas monkeys (*Erythrocebus patas*): A test of ecological models of group dispersion

LYNNE A. ISBELL & KARIN L. ENSTAM

Introduction

Competitive relationships result from competitive interactions over resources that can affect survival and reproduction. For male mammals, prospective mates are one such resource. For female mammals, food is more important than mates as a contestable resource (Trivers 1972). Its importance has been recognized by three ecological models that invoke either food distribution (van Schaik 1989, Wrangham 1980) or both food distribution and abundance (Isbell 1991) to explain variation in competitive relationships within and between groups of female primates (see also Isbell and Van Vuren 1996, Sterck *et al.* 1997, van Hooff and van Schaik 1992, Wrangham 1987). For both sexes, survival is also affected by other factors, such as disease and predation, and it is possible that individuals also compete for resources that in some way minimize exposure to disease or predators. In fact, one of the three models (here called the 'predation hypothesis') considers predation to be more important than food in its effects on the grouping behavior of females (van Schaik 1989).

The predation hypothesis assumes that predation ultimately forces females to live together and that variation in predation pressure causes variation in spatial cohesion within groups and among species (Sterck *et al.* 1997, van Hooff and van Schaik 1992, van Schaik 1989). Where predation pressure is high, females are predicted to decrease interindividual distances and thus live in groups that are spatially cohesive. The advantages of living closer to other group members to reduce predation may be gained if one places others between oneself and the predator (Hamilton 1971), or if the predator's ability to target particular individuals during an attack is reduced (Pulliam and Caraco 1984), or if more neighbors reduce detection time (van Schaik *et al.* 1983). Close proximity to others may increase food competition, however, and group members are expected to increase interindividual distances if given the opportunity. One such opportunity might arise when the risk of predation is low. Under low predation pressure, females are predicted to increase interindividual distances, either uniformly or between subgroups (fission–fusion grouping), thereby living less cohesively. In some species, dispersion is greatest while animals are foraging (Boinski *et al.* 2000). This has been interpreted as enhancing foraging efficiency at the cost of increasing vulnerability to predation (Boinski *et al.* 2000). The predation hypothesis thus implies that trade-offs exist between foraging efficiency and vulnerability to predation and that primates decrease foraging efficiency in exchange for greater safety from predators.

Contrasted with this model are two models that, although differing in the hypothesized effects of food on competition between groups, are similar in hypothesizing that food distribution is sufficient to explain variation in group dispersion (Isbell 1991, Wrangham 1980). The 'food distribution' hypothesis predicts that, where foods are spatially clumped, females decrease interindividual distances, and where foods are more spatially dispersed, females increase interindividual distances. The food distribution hypothesis also implies that if there are trade-offs between foraging and risk of predation, animals will maintain foraging efficiency at the risk of being more vulnerable to predators.

The predation hypothesis and the food distribution hypothesis present mutually exclusive predictions that can be tested under field conditions provided one factor varies while the other does not. For example, one might compare two populations of one species whose foods are distributed similarly but that live in habitats that differ in predation pressure. Alternatively, one might compare two closely related species whose food differs in distribution but which live in habitats with similar predation pressure. Comparison of closely related species reduces the chances that any differences could be explained by phylogenetic inertia.

Vervets (*Cercopithecus aethiops*) and patas monkeys (*Erythrocebus patas*) are two species that allow the latter comparison. Vervets and patas monkeys are more closely related to one another than they are to other primates (Disotell 1996), thus enabling us to exclude phylogenetic inertia as an explanation for observed behavioral differences between the two species. Vervets and patas monkeys are also sympatric in parts of their biogeographic ranges and overlap in body size (female vervets weigh 2.5–5.3 kg and female patas, 4.0–7.5 kg; Haltenorth and Diller 1977, Turner *et al.* 1997), making them vulnerable to the same predators in areas of sympatry. Although vervet groups typically have multiple males whereas patas groups have single males (most of the time), males in both species are similar to females in their responses to predators and in both species, males and females avoid predators more often than confront them (Cheney and Seyfarth 1981, Cheney and Wrangham 1987, Chism *et al.* 1983, LAI, unpublished data).

One area of sympatry is in Laikipia, Kenya, where a long-term comparative study of vervets and patas monkeys was initiated in 1992. At this study site, the food trees of vervets are more spatially clumped than the food trees of patas monkeys (Pruetz and Isbell 2000). In addition, the distance between food sites is greater for patas than for vervets (a food site is defined as any location where an animal stops to feed and that is separated from other food sites by hindlimb movement of the animal; Isbell *et al.* 1998). The home range of the study group of vervets can be further separated into two habitat types that also differ in food distribution at the scale of individual trees. Food trees in the part of their home range that includes *Acacia xanthophloea* riverine habitat are more clumped than food trees in the part of their home range that includes *A. drepanolobium* habitat (Pruetz and Isbell 2000) (the distance between food sites does not differ; Isbell *et al.* 1998). Comparison of the same group in these two habitats decisively eliminates any potential confounding influences of phylogenetic history, body size, group size, individual differences, and predation pressure, while allowing food tree distribution to vary.

Here we test opposing predictions generated by the predation and food competition models by examining group dispersion in sympatric vervets and patas monkeys sharing the same guild of predator species. Group dispersion is measured here by group spreads and interindividual distances. If the predation model is correct, there should be no significant difference between vervets and patas in group dispersion because both species are vulnerable to the same predators. Similarly, among vervets, there should be no significant difference in group dispersion between habitats within the group's home range because their predators are not constrained to one or the other habitat type. In addition, for individuals reducing foraging efficiency in return for greater safety from predators, interindividual distances are expected to be closer than interfood distances.

If, on the other hand, the food competition models are correct, vervets should be less dispersed than patas because their food trees are more clumped than are the food trees of patas. Vervets should also be less dispersed in that part of their home range where food trees are clumped and more dispersed in that part of their home range where food trees are less clumped. In addition, for individuals either not making trade-offs between foraging efficiency and vulnerability to predators or maintaining foraging efficiency at a risk of increasing their vulnerability to predators, interindividual distances are expected to be similar to interfood distances. We focus on females because the models were developed largely to explain variation in female grouping behavior.

Methods

Study site and subjects

The study was conducted as part of an ongoing comparative project begun in 1992 at Segera Ranch (36° 50′ E, 0° 15′ N; elevation 1800 m) on the Laikipia Plateau in central Kenya. The ecosystem is semi-arid, with mean annual rainfall of approximately 700 mm, although this varies considerably from year to year. Segera Ranch is a privately owned cattle ranch and conservation area that supports a wide diversity of wild animals, including most of the potential predators of vervets and patas monkeys, including leopards (*Panthera pardus*), lions (*P. leo*), black-backed jackals (*Canis mesomelas*), and martial eagles (*Polemaetus bellicosus*) (Table 10.1). Two major habitat types occur in the study area. Riverine areas support woodlands dominated by *Acacia xanthophloea* but that includes a smaller woody shrub layer (*Carissa edulis*, *Euclea divinorum*). Away from streams and rivers, vertisolic soils of impeded drainage ('black cotton soil') (Ahn and Geiger 1987) support woodlands dominated by *A. drepanolobium* and several species of grasses (predominantly *Pennisetum mezianum*, *P. stramineum*, and *Themeda triandra*) (Young *et al.* 1997). The two *Acacia* species differ considerably in height and canopy volume. While *A. xanthophloea* can grow to 25 m or more (Coe and Beentje 1991), *A. drepanolobium* only rarely grows to 7 m; 98% of individuals are 4 m or less (Isbell 1998, Young *et al.* 1997).

The behavioral data come from one group of vervets averaging 18.3 individuals (seven adult females and two adult males) and one group of patas monkeys averaging 24.2 individuals (nine to ten adult females and one adult male), all of whom were habituated to the presence of observers. All animals were individually identified

Table 10.1. *Signs of potential predators from November 1997–August 1999 in the home ranges of the study groups of vervets and patas monkeys. For each potential predator, the number of sightings by observers is given first, followed by the number of sightings of tracks and dung and reliable sightings of predators by cattle herders. The presence of nocturnal predators, such as hyenas, lions, and leopards, is more often determined by signs than by actual sightings whereas the presence of strictly diurnal predators, such as cheetahs, and martial eagles, is more often determined by sightings*

Species	Vervets	Patas
Lion (*Panthera leo*)	1/3	3/17
Leopard (*P. pardus*)	2/4	0/0
Cheetah (*Acinonyx jubatus*)	4/1	2/0
Spotted hyena (*Crocuta crocuta*)	0/4	0/3
Black-backed jackal (*Canis mesomelas*)	3/0	91/1
African wild cat (*Felis libyca*)	1/0	10/0
Serval (*F. serval*)	3/0	0/0
Caracal (*F. caracal*)	0/0	2/0
Martial eagle (*Polemaetus bellicosus*)	2/0	3/0

by natural markings and physical characteristics or by hair dye sprayed on their pelage with a syringe.

The vervet group lives along the Mutara River and defends its home range against incursions by neighboring groups. The vervets sleep and forage in *A. xanthophloea* habitat but also forage in adjacent *A. drepanolobium* habitat. Food trees in *A. xanthophloea* habitat are more spatially clumped than those in *A. drepanolobium* habitat (Pruetz and Isbell 2000). In addition, on average, in *A. xanthophloea* habitat, *A. xanthophloea* trees are 13.3 m from other *A. xanthophloea* trees and food sites are 5.8 m from other food sites, whereas in *A. drepanolobium* habitat, *A. drepanolobium* trees are 2.4 m from other *A. drepanolobium* trees and food sites are 6.2 m from other food sites for the vervet group (Isbell *et al.* 1998, Pruetz 1999). The divergence between inter-tree distance and inter-food site distance in these two habitats likely reflects the fact that individual *A. xanthophloea* are large, have multiple food sites within them, and are seldom passed without being fed in whereas individual *A. drepanolobium* trees are small, have few food sites, and are often passed without being fed in as vervets forage. With both habitat types combined, distances between trees and between food sites are 7.8 m and 6.1 m, respectively, in the home range of the vervet group (Isbell *et al.* 1998, Pruetz 1999). The patas group is restricted to *A. drepanolobium* habitat. On average, trees are 4.3 m from other trees and food sites are 16.5 m from other food sites in the home range of the patas monkey group (Isbell *et al.* 1998, Pruetz 1999), again reflecting the fact that not all trees are fed in as monkeys forage.

Data collection

Demographic data (births, deaths, disappearances, emigrations, and immigrations), dominance interactions, alarm calls, and predator sightings in conjunction with alarm calls have been recorded regularly since 1992 (number of observation days per month: patas monkeys, mean=7.4, mode=5, range=0–24; vervets: mean=7.4, mode=10, range=0–18). All predator sightings and signs within the home ranges of the monkeys, including those not associated with alarm calls, have been recorded since November 1997. From January to August, 1999, data on interindividual distances were collected from all adults using focal animal sampling. Focal animals were sampled beginning on the hour for 30 minutes using a predetermined random sampling procedure without replacement. Point samples were taken every 5 minutes during the 30-minute sample, with the identities of the three nearest neighbors of any age and of either sex and their distances from the focal female recorded. When the focal animal was 50 m or more away from any other animals, it was considered separated from the group, and alone or peripheral.

Inter-individual distances were calculated for all adult males and females in both groups (vervets: $n = 297$ point samples, range, 12–49 per individual; patas: $n = 481$ point samples, range 24–79 per individual), with the exception of one patas monkey (MIC), who was excluded because she died after being sampled only five times. The last two months of data on interindividual distances of vervets were excluded because the study group fused with another group in July, 1999 and the sudden and unusual addition of strange females to the group could have created a new group that was atypical in dispersion.

Data on group spreads were recorded from January to August, 1999 (data from July and August were excluded from analyses for vervets; see above), once every observation hour on the three-quarter-hour. While one observer stood at one edge of the group, another walked to the farthest visible edge and estimated the distance between himself and the other observer. Distances were always estimated by the same observer, and were consistent with estimates from previous studies (e.g., Isbell *et al.* 1998, 1999). Habitat type was recorded for a subset of these group spreads for vervets as part of another study (Enstam, unpublished data). No obvious bias could be detected with this subset.

Data analysis

Because female vervets and patas monkeys typically remain in their natal groups throughout life, permanent disappearances of adult females were considered deaths. The criteria for determining cause of death follow Cheney *et al.* (1988) as modified by Isbell (1990). Females that disappeared were considered to have died of suspected predation when they were in apparently good health within 72 h preceding their noted disappearance. Predation was confirmed if the predator was observed feeding on the monkey or if remains were found that could be assigned to a missing individual.

The data on group spreads and interindividual distances were entered into Excel (Microsoft) and then imported to JMP (SAS Institute, Cary, SC) for analysis. Analyses were conducted on mean interindividual distances per focal animal for each of the first, second, and third nearest neighbors.

With all data points on nearest neighbors of adult females included, the mean distance for the third nearest neighbor was actually smaller (9.3 m) than the mean for the second nearest neighbor (10.1 m) for vervets. This occurred because in several cases (QSO: 5; SAL: 6), the second nearest neighbor was less than 50 m from the focal animal while the third closest neighbor was greater than 50 m away from the focal animal and therefore not counted as a neighbor. By definition, however, the second nearest neighbor is always closer than the third. We therefore excluded those data points and recalculated the means so that the third nearest neighbor was indeed farther away, on average, than the second nearest neighbor.

Results

Predator presence

From November 1997 to August 1999, nine potential predator species of vervets and patas monkeys were observed either directly or via signs at least once, and six of these occurred in the home ranges of both study groups (Table 10.1). Although leopards

(*Panthera pardus*) were not seen during this time in the home range of the patas monkeys, they had been seen there in the past. Servals (*Felis serval*) were seen only in the vervets' home range, whereas caracals (*F. caracal*) were seen only in the patas monkeys' home range. These congeners are similar in body size and general diets, with caracals replacing servals in drier habitats (Dorst and Dandelot 1969, Estes 1991). Given that the home ranges of the two study groups are separated by only about 4 km, the overlap in predator species is not surprising. Although an analysis of densities of individual predator species cannot be done here, the number of sightings suggest that black-backed jackals, lions, and African wild cats (*F. libyca*) were more common in the habitat of patas monkeys, whereas leopards were more common in the habitat of vervets during this study. Only martial eagles and leopards are confirmed predators of vervets and only jackals are confirmed predators of patas at this site. It is unlikely that hyenas (*Crocuta crocuta*) are actual predators of primates because these nocturnal carnivores do not climb trees. With the exception of martial eagles and cheetahs, all of the potential predators hunt mainly at night.

Predation on adult females

Since the long-term study began, a minimum of ten of 18 (56%) adult female vervets and five of 34 (15%) adult female patas monkeys have died of suspected or confirmed predation. Remains of five of the ten adult female vervets were found, and signs of leopards near the group's locations around the dates of death suggest that most of these females were killed by leopards (see also Isbell 1990). In one episode of predation in which three vervets died overnight, claw marks were seen on the trunks of the vervets' sleeping trees, and remains of the vervets in and below the trees were found (V. Cummins and S. Robbins, personal communication). No remains of adult female patas monkeys that were suspected of being killed by predators have ever been found.

Interindividual distances

Since vervets were observed more often in early morning, and patas monkeys in mid-morning to late afternoon, we examined the possibility that the data could be biased by time of day before conducting further analyses. Interindividual distance was not correlated with time of day in either species, however (patas, first nearest neighbor: $r^2 = 0.44$, $p = 0.15$, df $= 48$; second nearest neighbor: $r^2 = 0.02$, $p = 0.35$; third nearest neighbor: $r^2 = 0.04$, $p = 0.19$; vervets, first

Fig. 10.1. Distances in meters (±1 SE) between the focal female and her three nearest neighbors. Nearest neighbors are significantly closer ($p = 0.001$) in vervets than in patas monkeys.

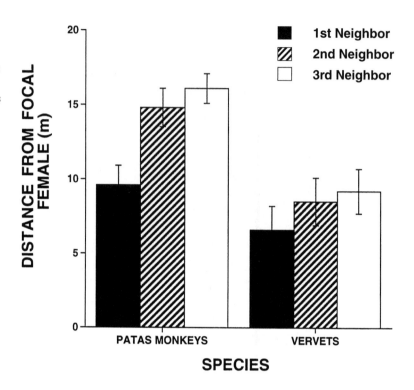

nearest neighbor: $r^2 = 0.0002$, $p = 0.93$; df = 43; second nearest neighbor: $r^2 = 0.02$, $p = 0.31$; third nearest neighbor: $r^2 = 0.02$, $p = 0.32$), suggesting that such a bias was not present.

The nearest neighbor of focal female patas monkeys was, on average, 45% farther away than was the nearest neighbor of focal female vervets (patas: 9.6 m ± 1.3 standard error; vervets: 6.6 m ± 1.6; t-test: $t = 1.43$, $p = 0.17$, df = 14; Fig. 10.1). The second nearest neighbor of focal female patas was 74% farther away than that of vervets (patas: 14.8 m ± 1.3; vervets: 8.5 m ± 1.6; $t = 3.04$, $p = 0.009$; Fig. 10.1). The third nearest neighbor of focal patas was 75% farther away than that of vervets (patas: 16.1 m ± 1.0; vervets: 9.2 m ± 1.5; $t = 4.03$, $p = 0.001$; Fig. 10.1). Together, the three nearest neighbors of adult female patas monkeys were significantly farther away from focal females than were those of adult female vervets (Fisher's combined test: $\chi^2 = 26.41$, $p < 0.001$, df = 6). Interindividual distances could not be analyzed for vervets in the two different habitats because sample sizes per female were too small.

Greater interindividual distances in patas monkeys relative to vervets were mirrored by greater distances between food sites in the home range of patas. The mean distance between food sites was 16.5 m for patas and 6.1 m for vervets (Isbell *et al.* 1998). The average

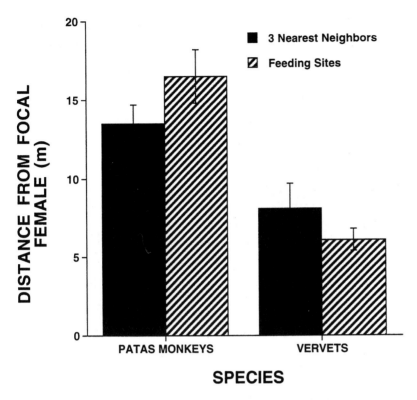

Fig. 10.2. Mean inter-individual distances of the three nearest neighbors of focal animals relative to mean inter-food site distances for vervets and patas monkeys. Inter-food site distances are from Isbell *et al.* 1998.

distance of the three nearest neighbors combined corresponded remarkably well with average interfood sites for both species (Fig. 10.2).

Group spreads

Group spread was not correlated with time of day (vervets, $r^2 = 0.03$, $p = 0.25$, $n = 48$; patas, $r^2 = 0.001$, $p = 0.73$, $n = 88$). The mean group spread for patas monkeys was 151.8 m ± 6.8 whereas the mean group spread for vervets was 128.0 m ± 10.2. Group spread was probably underestimated for patas monkeys. When group spread is estimated by multiplying the mean nearest-neighbor distance (9.6 m) by the mean number of animals in the group minus one (23.2), the group spread is 223 m, 47% greater than the estimate derived from observations in the field. Group spread probably was not underestimated for vervets. The mean nearest-neighbor distance of 6.6 m multiplied by the vervets' mean group size minus one (17.3) yields an estimated group spread of 114 m, only 11% off the estimate derived from observations in the field. The greater difficulty in locating individual patas monkeys for focal sampling (see Isbell *et al.* 1998) suggests that the difference in estimates of group spread

in patas monkeys occurred because patas monkeys are more difficult than vervets to see in the field. Even without adjusting for the
underestimation, vervets had significantly smaller group spreads
than patas ($t = 2.0$, $p = 0.05$, $n = 134$, two-tailed).

The greater group spread of patas was not caused by the resident
male. Although the resident male patas monkey may be socially
peripheral (Carlson 2000), he is apparently not spatially peripheral.
The nearest neighbor of the focal adult male patas monkey was, on
average, 58% farther away than the nearest neighbor of focal adult
male vervets (patas: 11.2 m; vervets: 7.1 m $\pm$ 0.8), but the difference
was not statistically significant ($t = 2.63$, $p = 0.23$). Assuming that the
resident male patas was always on the edge of the group, he would
have contributed, at most, about half the difference of 23.8 m
between species in group spread because he was, on average, 11.2 m
from his nearest neighbor.

When vervets used A. *xanthophloea* habitat, their mean group
spread was 93.0 m $\pm$ 4.6. When they used A. *drepanolobium* habitat,
their mean group spread was 165.8 m $\pm$ 18.2, 78% wider than in A.
xanthophloea habitat ($t = 3.54$, $p = 0.006$, $n = 11$). The wider group
spread of vervets in A. *drepanolobium* habitat is consistent with other
findings that vervets converge to some extent with patas monkeys
in foraging and ranging behavior when they use the same habitat
(Isbell *et al.* 1998).

Discussion

Vervets have been described as living in 'compact' groups (van Hooff
and van Schaik 1992: 362), whereas patas monkeys have been
described as living in less cohesive groups in which group members
are widely dispersed and 'separated by tens of meters most of the
time' (van Hooff and van Schaik 1992: 364). While this study does
not confirm the magnitude of separation within patas groups, the
data, none the less, support the point that vervets are generally less
dispersed than patas monkeys.

The estimates of group spreads and interindividual distances do
not support the hypothesis that the differences in dispersion
between vervets and patas monkeys are caused by differences in predation. According to the predation hypothesis, the smaller group
spreads and interindividual distances of vervets occur because
vervets are at high risk of predation and patas are at low risk. At this
study site, however, neither species can be argued to be at low risk
of predation. Females in both species died of predation, and the

same predators occurred in the home ranges of both vervets and patas. The more numerous items of hard evidence of predation in vervets, that is, remains of monkeys, could perhaps be taken to suggest that vervets suffer greater predation than patas. However, the larger size of the home range of the patas group (100 times the size of the home range of the vervet group) makes it much more difficult both to find the remains of patas monkeys that have disappeared and to find the group within the 72-h time limit for categorizing an apparently healthy female's disappearance as suspected predation. In support of the latter statement, the patas group was found most frequently on 5 days per month, not sufficiently often to determine probable cause of death for most females. In contrast, the vervet group was found most frequently on 10 days per month, enough to be within the 72-h time limit for classifying disappearances of apparently healthy females as suspected predation. With twice the annual mortality rate of female vervets (Isbell and Young, unpublished data), it is even possible that female patas suffer greater predation than female vervets.

The predation hypothesis is further undermined by the differences in dispersion in the same group of vervets using two different habitat types within their home range. It is difficult to argue that predation pressure differs significantly within the 40-ha home range of the vervet group during the day when the vervets are foraging. Predators typically have larger home ranges than their prey and, although two of the potential predators of vervets (servals and caracals) are apparently more constrained than the other predators to one or the other habitat type (Table 10.1), they do not leave a void but instead replace each other. There is no indication that the riverine habitat presents a barrier to other predators.

The comparison between species in group spreads and interindividual distances is, however, consistent with the food distribution hypothesis. Food trees of vervets in *A. xanthophloea* habitat are more clumped than foods of both vervets and patas in *A. drepanolobium* habitat (Pruetz and Isbell 2000). Moreover, distances between food sites are shorter in the home range of the vervets than in the home range of the patas monkeys. As predicted by the food distribution models, vervets were more spatially cohesive than patas, and more spatially cohesive in *A. xanthophloea* habitat than in *A. drepanolobium* habitat. Furthermore, distances between neighbors mirrored quite well the distances between food sites for both species.

Animals often have greater interindividual distances while foraging than while engaged in other activities (e.g., spider monkeys

(*Ateles belzebuth*): Klein and Klein 1975, squirrel monkeys (*Saimiri oerstedi*): Boinski 1987). The assumption is that they spread out to reduce competition with each other, even to the extent of increasing their vulnerability to predators (Boinski *et al.* 2000). We know of no studies that have actually examined food intake of individuals at different interindividual distances, however. In the absence of such data, the alternative explanation, that foraging requires the mapping of interindividual distances onto inter-food distances whereas other activities do not, cannot be ignored (see also Phillips 1995).

To conclude, we suggest that vervets and patas do not need to be particularly sensitive to predators as they forage because they feed during the day when most of their predators, including leopards, their most deadly predator, are least active.

Acknowledgments

We would like to thank L. Miller for inviting us to contribute this paper. The research was carried out with permission from the Office of the President, Republic of Kenya, and we thank our local sponsor, the Institute of Primate Research, and especially J. Mwenda, for facilitating logistics of conducting research in Kenya. We are grateful for the logistic support from our hosts on Segera Ranch, owners J. Ruggieri and J. Gleason, and manager P. Valentine, and for field assistance from A. and R. Carlson, R. Chancellor, M. Evans, R. Mohammed, and B. Musyoka Nzuma. The research was supported by funding from the NSF (BCS 9903949 and doctoral dissertation improvement grant SBR 9710514 for KLE), California Regional Primate Research Center (through NIH grant RR-00169), UC Davis Bridge Grant Program, and the UC Davis Faculty Research Grant Program to LAI, and grants from the LSB Leakey Foundation and the Wenner-Gren Foundation for Anthropological Research to KLE. Constructive criticism by L. Miller, S. Miller and P. Garber on a previous draft improved the final version considerably, and we thank them for their efforts.

REFERENCES

Ahn, P.M., and Geiger, L.C. (1987). *Kenya Soil Survey – Soils of Laikipia District.* Ministry of Agriculture, National Agricultural Laboratories, Kabete, Kenya.

Boinski, S. (1987). Habitat use by squirrel monkeys (*Saimiri oerstedi*) in Costa Rica. *Folia Primatologica* **49**: 151–67.

Boinski, S., Treves, A., and Chapman, C.A. (2000). A critical evaluation of the influence of predators on primates: Effects on group travel. In: S. Boinski and P.A. Garber, eds., *On the Move: How and Why Animals Travel in Groups*, Chicago: University of Chicago Press, pp. 43-72.

Carlson, A.A. (2000). Social Relationships and Mating Patterns of Patas Monkeys (*Erythrocebus patas*). Ph.D. dissertation, University of Wisconsin.

Cheney, D.L., and Seyfarth, R.M. (1981). Selective forces affecting the predator alarm calles of vervet monkeys. *Behaviour* **76**: 25-61.

Cheney, D.L., and Wrangham, R.W. (1987). Predation. In: B.B. Smuts, D.L. Cheney, R.M. Seyfarth, R.W. Wrangham and T.T. Struhsaker, eds., *Primate Societies*. Chicago: University of Chicago Press, pp. 227-39.

Cheney, D.L., Seyfarth, R.M., Andelman, S.J., and Lee, P.C. (1988). Reproductive success in vervet monkeys. In: T.H. Clutton-Brock (ed.), *Reproductive Success*, University of Chicago Press, Chicago, pp. 384-402.

Chism, J., Olson, D.K., and Rowell, T.E. (1983). Diurnal births and perinatal behavior among wild patas monkeys: Evidence of an adaptive pattern. *International Journal of Primatology* **4**: 167-84.

Coe, M., and Beentje, H. (1991). *A Field Guide to the Acacias of Kenya*. Oxford: Oxford University Press.

Disotell, T.R. (1996). The phylogeny of Old World monkeys. *Evolutionary Anthropology* **5**: 18-24

Dorst, J., and Dandelot, P. (1969). *A Field Guide to the Larger Mammals of Africa*. Boston: Houghton Mifflin.

Estes, R.D. (1991). *The Behavior Guide to African Mammals*. Berkeley: University of California Press.

Haltenorth, T., and Diller, H. (1977). *A Field Guide to the Mammals of Africa, Including Madagascar*. London: Collins.

Hamilton, W.D. (1971). Geometry for the selfish herd. *Journal of Theoretical Biology* **341**: 295-311.

Isbell, L.A. (1990). Sudden short-term increase in mortality of vervet monkeys (*Cercopithecus aethiops*) due to leopard predation in Amboseli National Park, Kenya. *American Journal of Primatology* **21**: 41-52.

Isbell, L.A. (1991). Contest and scramble competition: Patterns of female aggression and ranging behavior among primates. *Behavioral Ecology* **2**: 143-55.

Isbell, L.A. (1998). Diet for a small primate: Insectivory and gummivory in the (large) patas monkey (*Erythrocebus patas pyrrhonotus*). *American Journal of Primatology* **45**: 381-98.

Isbell, L.A., and Van Vuren, D. (1996). Differential costs of locational and social dispersal and their consequences for female group-living primates. *Behaviour* **133**: 1-36.

Isbell, L.A., Pruetz, J.D., and Young, T.P. (1998). Movements of vervets (*Cercopithecus aethiops*) and patas monkeys (*Erythrocebus patas*) as estimators of food resource size, density, and distribution. *Behavioral Ecology and Sociobiology* **42**: 123-33.

Isbell, L.A., Pruetz, J.D., Nzuma, B.M., and Young, T.P. (1999). Comparing measures of travel distance in primates: Methodological considerations and socioecological implications. *American Journal of Primatology* **48**: 87–98.

Klein, L.L., and Klein, D.J. (1975). Social and ecological contrasts between four taxa of Neotropical primates. In: R.H. Tuttle, ed., *Socioecology and Psychology of Primates*, The Hague: Mouton, pp. 59–85.

Phillips, K.A. (1995). Resource patch size and flexible foraging in white-faced capuchins (*Cebus capucinus*). *International Journal of Primatology* **16**: 509–19.

Pruetz, J.D. (1999). Socioecology of adult female vervet (*Cercopithecus aethiops*) and patas monkeys (*Erythrocebus patas*) in Kenya: Food availability, feeding competition, and dominance relationships. Ph.D. thesis, University of Illinois, Champaign-Urbana.

Pruetz, J.D., and Isbell, L.A. (2000). Correlations of food distribution and patch size with agonistic interactions in female vervets (*Chlorocebus aethiops*) and patas monkeys (*Erythrocebus patas*) living in simple habitats. *Behavioral Ecology and Sociobiology* **49**: 38–47.

Pulliam, H.R., and Caraco, T. (1984). Living in groups: Is there an optimal group size? In: J.R. Krebs and N.B. Davies, eds., *Behavioural Ecology: An Evolutionary Approach*. Sunderland, MA: Sinauer Associates, pp 122–47.

Sterck, E.H.M., Watts, D., and van Schaik, C.P. (1997). The evolution of female social relationships in nonhuman primates. *Behavioral Ecology and Sociobiology* **41**: 291–309.

Trivers, R.L. (1972). Parental investment and sexual selection. In: B. Campbell, ed., *Sexual Selection and the Descent of Man, 1871–1971*, Chicago: Aldine, pp. 136–79.

Turner, T.R., Anapol, F., and Jolly, C.J. (1997). Growth, development, and sexual dimorphism in vervet monkeys (*Cercopithecus aethiops*) at four sites in Kenya. *American Journal of Physical Anthropology* **103**: 19–35.

van Hooff, J.A.R.A.M., and van Schaik, C.P. (1992). Cooperation in competition: The ecology of primate bonds. In: A.H. Harcourt and F.B.M. de Waal, eds., *Coalitions and Alliances in Humans and Other Animals*, New York: Oxford University Press, pp. 357–89.

van Schaik, C.P. (1989). The ecology of social relationships amongst female primates. In: V. Standen and R.A. Foley, eds., *Comparative Socioecology: The Behavioural Ecology of Humans and Other Mammals*, Oxford: Blackwell, pp. 195–218.

van Schaik, C.P., van Noordwijk, M.A., Warsong, B., and Sutriono, E. (1983). Party size and early detection of predators in Sumatran forest primates. *Primates* **24**: 211–21.

Wrangham, R.W. (1980). An ecological model of female-bonded primate groups. *Behaviour* **75**: 262–300.

Wrangham R.W. (1987). Evolution of social structure. In: B.B. Smuts, D.L. Cheney, R.M. Seyfarth, R.W. Wrangham and T.T. Struhsaker, eds., *Primate Societies*. Chicago: University of Chicago Press, pp. 282–96.

Young, T.P., Stubblefield, C.H., and Isbell, L.A. (1997). Ants on swollen-thorn acacias: Species coexistence in a simple system. *Oecologia* **109**: 98–107.

11 • Predation risk and antipredator adaptations in white-faced sakis, *Pithecia pithecia*

TERRENCE M. GLEASON & MARILYN A. NORCONK

Introduction

The risk of predation poses a constant threat to the lives of primates living in natural habitats, and the study of its influence on many aspects of primate life has a long legacy in the history of primatology (Crook and Gartlan 1967, Eisenberg *et al.* 1972, Hart 2000, Kummer 1967, Terborgh 1983, Terborgh and Janson 1986). While these studies have been largely theoretical in nature, it seems clear that a number of different biological, ecological, and behavioral variables interact to constitute a given species' response to the threat of predation. Thus, for example, large body size may reduce the number of potential predators, a species may avoid areas where the density of predators (i.e., risk) is high, or small-bodied species may adopt a cryptic strategy in an effort to escape detection. Socially, individuals may give alarm calls to warn other members of the group in the event of danger, and/or manipulate their spatial proximity to other group members under different risk regimes (i.e., area or conditions of high vs. low risk, see Ydenberg 1998).

While problems exist in the interpretation of how these adaptations have evolved in concert with other social characteristics, it is difficult to imagine that the threat of being eaten represents anything but a strong selective force in the lives of nonhuman primates. Indeed, it is difficult to conceive of an ecological variable more closely related to individual fitness than the threat of death. However, predation is rarely observed directly, and this fact has led some to suggest that it may be of little consequence for the evolution of social structure (e.g., Cheney and Wrangham 1987). It is true that primatologists are limited to secondary sources of data (playback experiments, estimates of vigilance, alarm calling) when studying predation, and that there have been very few studies conducted from the predator's point of view (e.g., Emmons 1987, Rettig

1978, Wright *et al.* 1997, and see Sauther, Chapter 7 this volume). A recent comprehensive treatment of predator–prey relationships across primates incorporates salient variables such as predator behavior, body size, and habitat type (Hart 2000).

There are, however, sound reasons for thinking that secondary sources of data are sufficient, and in some cases even better than witnessing actual predation events. Students of predation have begun to realize that the persistent threat of predation may be more important than actual predation attempts themselves (for a review, see Lima and Dill 1990). Indeed, Lima (1986) has suggested that changes in individual behavior due to risk may have a larger impact on demography than actual deaths due to predation. Peckarsky *et al.* (1993) have shown that the perceived threat of predation can impact the activity, growth, and fecundity of animals in natural habitats. Experimental studies in the field and in the laboratory indicate that risk is such an important factor in the lives of foraging animals that they often behave as if a predator is constantly present (Lima and Dill 1990, Ydenberg 1998). Thus, sources of data, apart from actual predation events, have plenty to teach us about the importance of risk, even if we never see a predation attempt at all.

Attempts to assess the importance of predation as an organizing influence for social structure and social behavior across primates have met with mixed results. Boinski *et al.* (2000) have recently suggested that attempting to predict the evolutionary, ecological, and behavioral significance of predation by employing cross-species comparisons may be premature. They list a number of reasons for this view, including the paucity of reliable data devoid of preconceived notions about how predator and prey should behave (a predictable consequence of theory preceding the accumulation of data). In addition to this well supported point of view, we would argue that half of the equation pertinent to the question at hand has been largely ignored. Studies of the behavior and ecology of predators that threaten primates have lagged far behind both the collection of relevant primatological data and the generation of theory purported to use these data in a hypothetical framework. Examining half of the data relevant to predator–prey interactions between primates and their predators can only result in an incomplete understanding of these interactions and their influence on the lives of both predator and prey.

Despite the fact that they appear to exhibit a range of different antipredator adaptations, no data exist on the specifics of these

Table 11.1. *Potential saki monkey predators documented from this region of eastern Venezuela*

Species	Common name	Venezuelan name
Morphus guianensis	Crested Eagle	Águila Monera
Buteo albonotatus	Zone-tailed Hawk	Gavilán Negro
Heterospizias meridionalis	Savanna Hawk	Gavilán Pita Venado
Harpy haliaetus solitarius	Solitary Eagle	Águila Solitaria
Harpia harpyja	Harpy Eagle	Águila Harpia
Spizaetus ornatus	Ornate Hawk Eagle	Águila de Penacho
Leopardus wiedii	Margay	Gato Tigre
Leopardus pardalis	Ocelot	Tigrillo
Panthera onca	Jaguar	Tigre
Leopardus tigrinus	Oncilla	Tigrillo
Herpailurus yaguarondi	Jaguarundi	Yaguarundi
Eira barbara	Tayra	Guache, Guanico
Eunectes murinus	Green anaconda	Anaconda
Boa constrictor	Red-tailed boa	Boa

Source: Linares, 1998; Phelps and de Schauensee, 1978; L. Balbás personal communication)

patterns for any of the pitheciin monkeys (sakis and uakaris). At an average body weight of 1.77 kg ($n = 3$ adults, Glander and Norconk, unpublished data), white-faced sakis are potential prey for a number of avian and terrestrial predators (Table 11.1). The avian predators are large hawks and eagles. Among these only the harpy eagle (*Harpia harpyja*) has been studied extensively. Rettig (1978) noted that harpy eagles in Guyana routinely take large primates such as red howler monkeys and that white-faced saki remains were found beneath harpy nests. They perch high in the tallest trees of the canopy but ambush prey within the canopy and subcanopy. Harpy eagles may be locally extinct in the study area due to habitat fragmentation (Alvarez-Cordero, personal communication). The conservation status of the other raptor species in the area of the lake is completely unknown. We have observed the zone-tailed hawk (*Buteo albonotatus*) and the crested eagle (*Morphus guianensis*), although we often observe raptors without being able to identify them to species. Zone-tailed hawks inhabit savanna habitats and secondary forest edges feeding on reptiles, birds, and small mammals. Crested eagles soar high in search of potential food but also ambush prey in the canopy and subcanopy. They have been reported to prey on primates as large as young adult spider monkeys (Julliot 1994).

Undoubtedly, the jaguar (*Panthera onca*), ocelot (*Leopardus pardalis*), and the red-tailed boa (*Boa constrictor*) are the primary terrestrial threats to white-faced sakis. All of these species have been observed in the study area. Jaguars have been recorded preying on a group of red howlers on a nearby island (Peetz *et al.* 1992), and jaguar tracks, as well as those of other felids, are occasionally observed along exposed banks during the dry season (personal observation). In addition, because sakis have been observed going to the water's edge to drink during the dry season (Harrison 1998), the green anaconda (*Eunectes murinus*) must be considered a potential threat during this time of year.

Smaller felids and large mustelids may also pose a threat to juvenile or elderly individuals. Tayra (*Eira barbara*) have been locally observed, are accomplished climbers, and have been known to take juvenile primates (Defler 1980). Margay (*Leopardus wiedii*), jaguarundi (*Herpailurus yaguarondi*), and oncilla (*Leopardus tigrinus*) are small felids that have been observed in the lake region, and all are capable of taking small-bodied mammals. Felids and other terrestrial mammals are more often temporary than permanent residents on small Guri Lake islands. Sakis have never been seen to swim between islands, but most other vertebrates (including felids and snakes) have been observed swimming in the lake.

In this chapter we present field data on a suite of antipredator behaviors in wild white-faced sakis, *Pithecia pithecia*, and the consequences of these behaviors for foraging individuals. We use patterns of alarm calls, group responses to perceived predator risk, and data on intragroup spacing and habitat use in an effort to describe the particular way in which white-faced sakis respond to the threat of predation. We view such baseline, descriptive ecological data as a fundamental precursor to the generation of theoretical approaches to foraging and social behavior.

Methods

Study site

Our research takes place on islands in Guri Lake, Bolívar State, eastern Venezuela. Guri Lake is a catchment area for the Raúl Leoni Hydroelectric Dam. The lake covers an area of 4200 km^2 and contains more than 100 islands, many of which are completely

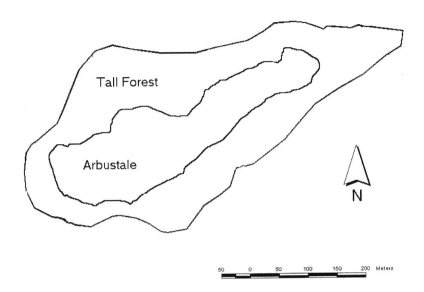

Fig. 11.1. Schematic map of the study island showing the distribution of the two major forest types, tall forest and low scrub forest, or arbustale.

forested. Flooding of the basin was completed in 1986 isolating the primates on the islands to which they had dispersed. Four primate species live in forests in and around the lake: *Pithecia pithecia, Chiropotes satanus, Alouatta seniculus,* and *Cebus olivaceus*. The present study was conducted on Isla Redonda (7° 45′N, 62° 52′W) in the northern section of the lake. The 15-ha island is composed of both tall, dry tropical forest and lower 'arbustale' forest in the center of the island (Fig. 11.1). In addition to white-faced sakis, the island is also inhabited by a single group of red howlers and several solitary howler individuals. Detailed descriptions of the island have been published elsewhere (Aymard *et al.* 1997, Norconk 1996, Parolin 1993). The climate is tropical with rainfall mediated by the Intertropical Convergence Zone, resulting in distinct wet (May–October) and dry (November–April) seasons. The study island receives an annual average rainfall of 1100 mm (EDELCA: Electrificación del Caroní, unpublished data).

Subjects

White-faced sakis are sexually dichromatic (males bearing white faces), with about 500 g difference in body weight between adult males and females (males larger than females). Group size during the years from which the present data are drawn has fluctuated between a maximum of nine individuals (five males, four females) and a minimum of five individuals (two males, three females). All individuals were fully habituated to human observation.

Fig. 11.2. Weighted index of centrality. The Centrality Index (CI) for each individual is equal to the sum of the proportion of samples occupying each of the four spatial zones, divided by the total number of samples. The zones in the numerator are according to theoretical optima. Zones A, B, and C represent 5-m concentric circles around a group centroid. Individuals distributed more than 15 m from the center are assigned to zone D.

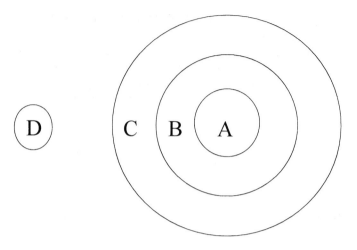

$$CI = \Sigma \; \frac{4nA + 3nB + 2nC + nD}{nA + nB + nC + nD}$$

Data collection

We present data collected during the years 1991–93, 1995, and 1997–99. Study periods varied in length from 3 months to 12 months. Alarm call data were collected as all-occurrence samples. We recorded the age and sex of the initial caller, whether or not the call was echoed, and how many times it was echoed. We also recorded the duration of the event and group response, that is, whether or not the sakis shifted position in the canopy (moved up or down) and whether the predator was mobbed or chased. We identified the predators as avian or terrestrial and identified them to species when possible.

Data on intragroup spacing were collected using both focal animal and group scan sampling methods (Altmann 1974) during the wet season of 1993 from 12 June to 5 September. A single focal animal was sampled for an entire day, and focal animals were rotated to ensure an even representation in the database. Every 20 minutes, we recorded the focal animal's activity (feed, rest, or travel), the type of food item taken when feeding, and the number and identity of other group members within 5 m (near neighbors). We also recorded the distance to each animal relative to the observer with a visual estimate of distance in meters and a direction taken as a compass bearing (Fig. 11.2). To avoid biasing the sample

Table 11.2. *Frequency and percent of total observations of alarm calls given to classes of predators*

Predator class	Frequency	Percentage of total observations	Number of species of predator
Vulture	174	77.0	2
Raptor	11	4.9	2
Snake	6	2.7	1
Felid	2	0.9	1
Unknown	33	14.6	—
Total	226	100.0	6

in favor of individuals that were easily seen, we discarded any samples that contained fewer than seven of the nine individuals in the group. A test of possible temporal autocorrelation that would bias the samples resulted in no significant change in the outcome of statistical analysis.

The Law of Cosines was applied to group scans *a posteriori* to calculate interindividual distance of animals relative to the focal animal:

$$c^2 = a^2 + b^2 - 2ab(\cos\alpha)$$

where a is the distance between the observer and individual (a), b is the distance between the observer and individual (b), and α is the angle between a and b. These individual estimates were plotted and used to derive a measure of each animal's relative position within the group using a Centrality Index (CI) following the method of Barton and Whiten (1993). All data were analyzed using SPSS v.10.

Results

Alarm calls
We recorded 226 separate occasions when white-faced sakis uttered alarm calls in response to predators (Table 11.2). Of these, 90.3% ($n = 204$) were directed at avian threats. The remainder (9.7%; $n = 22$) were directed at terrestrial predators. The potential predators were identified on 193 (85.4%) occasions, the vast majority of which (77%; $n = 174$) were common vultures. Crested eagles were spotted on three occasions and hawks on four occasions. The most common terrestrial threat was the red-tailed boa (3%; $n = 6$).

There was a strong sex bias in the tendency to utter alarm calls

($\chi^2 = 16.2$; df$=1$; $p < 0.01$) with males calling significantly more than expected. However, juveniles were also significantly more likely to utter alarm calls than adults ($\chi^2 = 23.4$; df$=1$; $p < 0.01$). In fact, the two juveniles in the group accounted for more than half of all alarm calls uttered (51.8%, 117 calls). Both of these trends are the result of the juvenile male, TX, being responsible for nearly 40% of all calls uttered.

Fifty-five percent of all calls were echoed by group members, and of these, 31% were echoed only once, 15% were echoed by other individuals twice, and 2.5% were echoed more than five times.

Group spacing

Vigilance behavior is difficult to measure in white-faced sakis due to their saltatory locomotion, as well as a high level of overall social vigilance (*sensu* Boinski 2000). As a result, it is difficult to tease apart a heightened sense of awareness due to feeding or social behavior/territorial behavior from vigilance related specifically to predation. However, if animals on the periphery of the group are more vigilant, they should: (a) be the first to give an alarm call when a predator is detected, and (b) give more alarm calls than animals in the core of the group. The animals with the three lowest centrality index scores (i.e., those most peripheralized) gave 68% of all initial alarm calls. There was no relationship between the total number of calls for each individual and that individual's position in the group.

We tested the effect of the median number of near neighbors (individuals within 5 m of the focal subject) on the tendency to utter alarm calls and found no significant association (Kendall's $\tau = 0.12$; $p > 0.05$; $n = 9$ individuals).

With respect to foraging success and predation risk, we examined whether the trends in centrality indices and/or near neighbor measures influenced an individual's foraging success. Only when feeding on insects was an individual's position within the group associated with its ability to access resources ($n = 9$, $\tau = 0.556$, $p < 0.05$). Sakis feed on insects rarely (opportunistically); hence, this association may be an artifact of small sample size. Next, we compared an individual's median number of near neighbors to the frequency with which it fed on each of the three food classes. Feeding frequencies did not vary with the number of other individuals within 5 m (fruit: $n = 9$, $\tau = -0.12$, $p > 0.05$; leaves: $n = 9$, $\tau = 0.29$, $p > 0.05$; insects: $n = 9$, $\rho = 0.34$, $p > 0.05$). Hence, neither group geometry (judged by centrality index scores), social distance (judged by nearest neighbor data), nor vigilance (judged by alarm calling) appear to have an effect on foraging success for individual white-faced sakis.

Group response

White-faced sakis generally maintain their position in the canopy (i.e., do not move lower or higher) under all but the strongest of threats (91.2%, $n = 206$ alarm bouts). However, strong threats (real or perceived) lead them to change their position in the canopy and change their foraging behavior in drastic fashion. Long bouts of alarm calling, or mobbing, directed at a strong threat occurred on four occasions. These bouts ranged in duration from 1.9 minutes to 88 minutes. The behavior of the sakis during these mobbing events provides insight into the overall white-faced saki response to strong threat of predation. For example, on 25/6/99 at 06:10 h, shortly after leaving their sleeping trees, all members of the group began alarming at a small felid (probably an oncilla). The entire group followed the predator over 200 m to the north side of the island, constantly alarming as they traveled. The mobbing lasted for over 20 minutes, after which the group climbed into several tall trees and remained motionless and extremely vigilant for 92 minutes. This incident resulted in their first feeding bout of the day being delayed nearly 2 h. On a second occasion, 18/12/91 the group alarmed for 20 minutes to a boa constrictor in a feeding tree.

Responses such as these are similar when confronted by a strong threat (real or perceived). For example, during the wet season of 1997, several howler monkeys died on the island from what may have been toxification from feeding on *Coccoloba striata* leaves (Polygonaceae). The death of the first individual attracted dozens of avian carrion feeders (including the 3.5 kg king vulture, *Sarcoramphus papa*) flying very low over the northeast end of the island. In contrast to the response to terrestrial threats, the sakis did not mob these potential threats; rather, they descended into the low, dense understory, spread themselves evenly over an area of about 1 ha, and remained motionless for 2 h and 13 min. Each individual remained very still and none uttered an alarm call, despite the presence of so many large birds flying low over the canopy. They eventually descended nearly to the ground and silently moved away, despite the continued presence of the vultures. In total, more than 3 h passed before the group resumed feeding. While this situation was extreme, it seems clear that sakis rarely mob avian predators unless they are very small hawks who have alighted in the canopy.

Habitat use

Figure 11.1 shows the distribution of the two primary habitat types found in the study area. Tall forest averages 18–20 m in height and provides cover from above and below. The predominant saki food species in this habitat are *Connarus venezuelanus* (Connaraceae), *Peltogyne floribunda* (Caesalpiniaceae) and *Licania discolor* (Chrysobalanaceae). These trees are among the tallest and most abundant trees on the island. Arbustale forest is low, scrubby, more xeric forest on extremely rocky soil. Important saki food sources in this habitat are *Erythroxylum steyermarkii* (Erythroxylaceae), *Alibertia latifolia* (Rubiaceae) and *Eugenia monticola* (Myrtaceae). These slender stemmed trees provide little cover. Thus given its physiognomy, arbustale provides little cover from avian or terrestrial threats, suggesting that white-faced sakis should be more vigilant and therefore utter more alarm calls in this habitat. Interestingly, sakis alarmed at a much higher rate (2.4 calls/h) in more protected, mature forest habitats than they did in more exposed arbustale habitats (0.8 calls/h).

Discussion

In this chapter we take the view that predation is a crucial aspect of the ecology of wild primates, and that comparable data need to be collected for the construction of theoretical approaches purporting predation as an organizing influence on many aspects of primate behavior and ecology. We have presented a preliminary analysis of alarm calling, group geometry, social spacing, and habitat use in Venezuelan white-faced sakis. We have also advocated an approach in which more attention must be paid to studying predation from the predator's point of view. Although strides in this area have been made using predator alarm call playback and stuffed predator experiments (Chapman and Chapman 1996, Cheney and Seyfarth 1981, 1985, 1990, Macedonia and Young 1991, van Schaik and van Noordwijk 1989), these studies still fall far short of the kind of information that would be provided by studies of the behavior and ecology of the predators themselves. Specifically, field data on how predators locate, pursue, attack, and retain primate prey would greatly expand our ability to formulate incisive questions about predation and its effect on the lives of primates (Kerfoot and Sih 1987). To counteract the strategies used by predators, prey species have two basic options: avoidance and deterrence (Brodie *et al.* 1991, Pulliam and Caraco 1984, Sih 1987). It is useful to examine the data presented here in the light of these two options. Our assumption is that which one of these strategies (or combination of strategies) sakis employ will have varying consequences for their foraging behavior.

Like most primates, white-faced sakis use alarm calls to warn other group members in the event of possible danger. Our data indicate that sakis perceive the most common risk to be from avian threats, which seems due to several factors. While both avian and terrestrial predators are probably equally rare, mistakes are rarely made in identifying terrestrial threats (primatologists and the odd lumbering tapir aside). In neotropical forests, terrestrial taxa large enough to be potential predators usually are actual predators, whereas large birds exhibit a much wider range of dietary strategies. This, coupled with the fact that avian predators may be more difficult to identify quickly, may in part explain this trend in sakis. A second possibility is that white-faced sakis are attacked far more often by avian predators than by terrestrial ones. Our data do not allow us to distinguish whether more attacks come from birds versus felids, snakes, and mustelids (actual predators: 11 avian, eight terrestrial); however, 33 avian threats were unidentified, suggesting that these may be slightly more common. Of course, in any particular habitat, the relative frequency of attacks will depend on the relative densities of the two types of predators, as well as the activity of those predators. This fact lends credence to the view that studies of the predators merit much further attention.

The proximate factors determining group geometry among terrestrial Old World monkeys have received wide attention, especially for baboons (Barton and Whiten 1993, Devore 1965, Rhine 1975). Individuals in the vanguard of moving troops or on the periphery of stationary groups are thought to experience enhanced predation risk (Hamilton 1971, Janson 1990). Vanguard individuals are the first to be exposed to predators in new areas of a given habitat (Boinski 2000), and peripheralized individuals suffer from a greater statistical likelihood of being attacked because they have fewer near neighbors (Hamilton 1971). Our results presented here are in accord with these predictions. Sakis with relatively lower centrality index scores produced significantly more alarm calls, and individuals who maintained shorter interindividual distances produced significantly fewer alarm calls. Few arboreal taxa have been studied in this light (but see Boinski 2000, Janson 1990), but the behavior of those that have seems to be well predicted by the models. One aspect of this research that may warrant future investigation is the possibility of expanding the models to include three dimensions. The original models were developed from the study of large herds and terrestrial baboons where individuals are threatened by attack in two dimensions, whereas arboreal taxa are threatened by attack in three dimensions.

Enhanced vigilance is thought to occur to the detriment of foraging success. Despite the fact that alarm calling varied predictably with group geometry and social distance, these differences did not translate into differential foraging success for individual sakis. It is possible that our chosen estimate of feeding (time spent feeding on each of three resource classes) does not provide the proper resolution for distinguishing fine-grained differences in foraging success. However, Norconk et al. (1999) used a fine-grained measure of intake (timed feeding events normalized by wet and dry food weights) to examine individual female foraging success given incidents of contest competition. They found that only when feeding on the rarest resources of unusually high nutrient quality could any differences in feeding success be discerned. If there are individual differences in feeding competition in white-faced sakis, our preliminary analyses suggest that it may be due to ecological reasons other than predation risk and/or social reasons other than contest competition.

Group responses to predation have been described for a number of platyrrhines. We described two different types of group response in Venezuelan white-faced sakis. When threatened by terrestrial threats, at least those posed by small felids and boas, sakis will mob and/or chase these predators through the forest. A period of extensive inactivity usually follows such events. Such mobbing events may be unusual for small-bodied primates living in small groups. Indeed, most species that are reported to mob and chase predators are large-bodied taxa living in large troops (chimpanzees: Hiraiwa-Hasegawa et al. 1986, baboons: Altmann and Altmann 1970, Iwamoto et al. 1996), though small cats are occasionally mobbed by smaller taxa (Passamani 1995).

Faced with strong avian threat, white-faced sakis tend to employ an extreme mode of cryptic evasion, whereby they descend to thick undergrowth and freeze for extended periods of time. This 'freezing' behavior has been described for a number of primate species (Gautier-Hion 1973, Izawa 1978, Wahome et al. 1993), most of them with relatively small body size.

As discussed above, we found that individual white-faced sakis vary little in their foraging success relative to several ecological and social variables. As a group, however, cryptic predator evasion can have a drastic effect on daily food intake. In the future, we hope to be able to compare pre-attack and post-attack foraging effort for individuals and for the group as a whole. We suspect that such lengthy periods of time with no feeding might be compensated for in several ways. First, sakis may feed at higher rates once feeding resumes. Second, they may concentrate on especially high-quality resources in the aftermath of a predator scare. Finally, daily activity budgets may be altered such that less time is devoted to nonfeeding activities such as rest, grooming, and play, and they may forage later into the afternoon.

Conclusion

White-faced sakis appear to employ a combination of detection and evasion to combat the threat of predation depending on the type of predator involved and the severity of the threat. Small terrestrial, arboreal, and perched avian predators, and snakes tend to evoke a mobbing response, whereas other strong avian threats lead to cryptic evasion. Unfortunately, we have no data on group response when faced with the most serious threats (harpy eagle and jaguar), though we suspect both would elicit an evasive response.

We found no evidence to suggest that different individual risk regimes lead to variable foraging success. However, extended periods of cryptic evasion undoubtedly affect the foraging success of all individuals in a group. Future work is needed to understand how sakis recoup lost foraging effort following serious predator attacks.

Acknowledgments

We would like to thank Amy Harrison and Jason Brush for providing us with additional data for this analysis. TMG would also like to than D. Tab Rasmussen for his comments and many long discussions about primates and predators. We are particularly grateful for the long-term assistance of Estudios Básicos of EDELCA-Guri, headed by TSU Luis Balbás, and to CONICIT (Consejo Nacionál de Investigaciones Científicas y Tecnológicas) for permission to work in Venezuela. This research was supported by NSF BNS 90–20614 to W.G. Kinzey & MN, SBR 98–07516 to MN, REU supplement to BNS 90–20614 to TMG, and the Wenner Gren Foundation for Anthropological Research.

REFERENCES

Altmann, J. (1974). Observational studies of behavior: sampling methods. *Behaviour* **14**: 227–65.

Altmann, S.A., and Altmann, J. (1970). *Baboon Ecology.* Chicago: University of Chicago Press.

Aymard, G, Kinzey, W.G., and Norconk, M.A. (1997). Composición florística de bosques en islas en el embalse de Guri, baja Rio Caroní, Edo. Bolívar. *BioLlania Edición Especial* **6**: 195–233.

Barton, R.A., and Whiten, A. (1993). Feeding competition among female olive baboons. *Animal Behavior* **46**: 777–89.

Boinski, S. (2000). Social manipulation within and between groups mediates primate group movement. In: S. Boinski and P.A. Garber, eds., *On the move: How and Why Animals Travel in Groups,* Chicago: University of Chicago Press, pp. 421–69.

Boinski, S., Treves, A., and Chapman, C.A. (2000). A critical evaluation of the influence of predators on primates: effects on group travel. In: S. Boinski and P.A. Garber, eds., *On the Move: How and Why Animals Travel in Groups,* Chicago: Universtiy of Chicago Press, pp. 43–72.

Brodie, E.D., Formanowicz, D.R., and Brodie, E.D. (1991). Predator avoidance and anti-predator mechanisms: Distinct pathways to survival. *Ethology, Ecology, and Evolution.* **3**: 73–7.

Chapman, C.A., and Chapman, L.J. (1996). Mixed species primate groups in Kibale National Forest: Ecological constraints on association. *International Journal of Primatology* **17**: 31–50.

Cheney, D.L., and Seyfarth, R.M. (1981). Selective forces affecting the predator alarm calls of vervet monkeys. *Behaviour* **76**: 25–61.

Cheney, D.L., and Seyfarth, R.M. (1985). Vervet monkey alarm calls: manipulation through shared information. *Behaviour* **92**: 150–66.

Cheney, D.L., and Seyfarth, R.M. (1990). *How Monkeys See the World: Inside the Mind of Another Species.* Chicago: University of Chicago Press.

Cheney, D.L., and Wrangham, R.W. (1987). Predation. In: B.B. Smuts, D.L. Cheney, R.M. Seyfarth, R.W. Wrangham and T.T. Struhsaker, eds., *Primate Societies.* Chicago: University of Chicago Press, pp. 227–39.

Crook, J.H., and Gartlan, J.S. (1967). Evolution of primate societies. *Nature* **210**: 1200–3.

Defler, T.R. (1980). Notes on interactions between the tayra (*Eira barbara*) and the white-fronted capuchin (*Cebus albifrons*). *Journal of Mammalogy* **61**: 156.

Devore, I. (1965). *Primate Behavior: Field Studies of Monkeys and Apes.* New York: Holt, Rhinehart, and Winston.

Eisenberg, J.F., Muckenhirn, N.A., and Rudran, R. (1972). The relation between ecology and social structure in primates. *Science* **176**: 863–74.

Emmons, L. (1987). Comparative feeding ecology of felids in a neotropical rain forest. *Behavioral Ecology and Sociobiology* **20**: 271–83.

Gautier-Hion, A. (1973). Social and ecological features of talapoin monkeys: Comparisons with sympatric cercopithecines. In: R.P. Michael and J.H. Crook, eds., *Comparative Ecology and Behavior of Primates*, New York: Academic Press, pp. 148–70.

Hamilton, W.D. (1971). Geometry for the selfish herd. *Journal of Theoretical Biology* **31**: 295–311.

Harrison, A.L. (1998). Feeding party dynamics of white-faced sakis in Lago Guri, Venezuela. Master's Thesis, Kent State University.

Hart, D. (2000). Primates as prey: ecological, morphological, and behavioral relationships between primate species and their predators. Ph.D Dissertation , Washington University, St. Louis.

Hiraiwa-Hasegawa, M., Byrne, R.W., Takasaki, H., and Byrne, J.M.E. (1986). Aggression toward large carnivores by wild chimpanzees of Tanzania. *Folia Primatologica* **47**: 8–13.

Iwamoto, T.A., Mori, A., Kawai, M., and Bekele, A. (1996). Anti-predator behavior in gelada baboons. *Primates* **37**: 389–97.

Izawa, K. (1978). A field study of the ecology and behavior of the black-mantled tamarin (*Saguinus nigricollis*). *Primates* **19**: 241–74.

Janson, C.H. (1990). Ecological conseqences of individual spatial choice in foraging groups of brown capuchin monkeys (*Cebus apella*). *Animal Behaviour* **90**: 922–34.

Julliot C. (1994). Predation of a young spider monkey (*Ateles paniscus*) by a crested eagle (*Morphus guianensis*). *Folia Primatologica* **63**: 75–7.

Kerfoot, W.C., and Sih, A. (1987). Introduction. In: W.C. Kerfoot and A. Sih, eds., *Predation*, ii–viii. University Press of New England, Hanover.

Kummer, H. (1967). Dimensions of a comparative biology of groups. *American Journal of Physical Anthropology* **27**: 357–66.

Lima, S. (1986). Predation risk and unpredictable foraging conditions: determinants of body mass in birds. *Ecology* **67**: 377–85.

Lima, S., and Dill, L. (1990). Behavioral decisions made under risk of predation: a review and prospectus. *Canadian Journal of Zoology* **68**: 619–40.

Linares, O.J. (1998). *Mamíferos de Venezuela*. Caracas: Sociedad Conservacionista Audubon de Venezuela.

Macedonia, J.M., and Young, P.L. (1991). Auditory assessment of avian predator threat in semicaptive ringtailed lemurs (*Lemur catta*). *Primates* **32**: 169–82.

Norconk, M.A. 1996. Seasonal variation in the diets of white-faced and bearded sakis (*Pithecia pithecia* and *Chiropotes satanas*) in Guri Lake, Venezuela. In: M.A. Norconk, A.L. Rosenberger, P.A. Garber, eds., *Adaptive Radiations of Neotropical Primates*, New York: Plenum, pp. 403–23.

Norconk, M.A., Gleason, T.M., and Harrison, A.L. (1999). Feeding rates and social dominance among white-faced saki females. *American Journal of Physical Anthropology*, **Suppl. 28**: 212.

Parolin, P. 1993. Forest inventory in an island of Lake Guri, Venezuela. In: W. Barthlott, C.M. Naumann, K. Schidt-Loske, and K.L. Schuchmann, eds., *Animal–Plant Interactions in Tropical Environments*. Bonn: Museum Alexander Koenig, pp. 139–47.

Passamani, M. (1995). Field observations of a group of Geoffroy's marmosets mobbing a margay cat. *Folia Primatologica* **64**: 163–6.

Peckarsky, B.L., Cowan, C.A., Penton, M.A., and Anderson, C. (1993). Sublethal consequences of stream-dwelling predatory stoneflies on mayfly growth and fecundity. *Ecology* **74**: 1836–46.

Peetz, A., Norconk, M.A., and Kinzey, W.G. (1992). Predation by jaguar on howler monkeys (*Alouatta seniculus*) in Venezuela. *American Journal of Primatology* **28**: 223–8.

Phelps, W.H., and de Schauensee, R.M. (1978). *A Guide to the Birds of Venezuela*. Princeton: Princeton University Press.

Pulliam, H.R., and Caraco, T. (1984). Living in groups: Is there an optimal group size? In: J.R. Krebs and N. Davies, eds., *Behavioral Ecology*. Oxford: Blackwell Scientific Publishing, pp. 122–47.

Rettig, N.L. (1978). Breeding behavior of the harpy eagle (*Harpia harpyja*). *Auk* **95**: 629–43.

Rhine, R.J. (1975). The order of movement of yellow baboons (*Papio cyanocephalus*). *Folia Primatologica*, **23**: 72–104.

Sih, A. (1987). Predators and prey lifestyles: An evolutionary and ecological overview. In: W.C. Kerfort and A. Sih, eds., *Predation*, Hanover: University Press of New England, pp. 203–24.

Terborgh, J. (1983). *Five New World Primates: A Study in Comparative Ecology*. Princeton: Princeton University Press.

Terborgh, J., and Janson, C.H. (1986). Socioecology of primate groups. *Annual Review of Ecological Systematics* **17**: 111–35.

van Schaik, C.P., and van Noordwijk, M.A. (1989). The special role of male *Cebus* monkeys in predation avoidance and its effects on group composition. *Behavioral Ecology and Sociobiology* **24**: 265–76.

Wahome, J.M., Rowell, T.E., and Tsinglia, H.M. (1993). The natural history of the de Brazza's monkey in Kenya. *International Journal of Primatology* **14**: 445–66.

Wright, P.C., Heckscher, S.K., and Dunham, A.E. (1997). Predation on Milne-Edward's sifaka (*Propithecus diadema edwardsi*) by the fossa (*Cryptoprocta ferox*) in the rain forest of southeastern Madagascar. *Folia Primatologica* **68**: 34–43.

Ydenberg, R.C. 1998. Behavioral decisions about foraging and predator avoidance. In: R. Dukas, ed., *Cognitive Ecology: The Evolutionary Ecology of Information Processing and Decision Making*, Chicago: University of Chicago Press, pp. 343–70.

Part III • Environmental variables

12 • Foraging female baboons exhibit similar patterns of antipredator vigilance across two populations

RUSSELL A. HILL & GUY COWLISHAW

Introduction

Predation pressure has long been considered a powerful selective force on primate behavior (Alexander 1974, Anderson 1986, Crook and Gartlan 1966, Dunbar 1988, Isbell 1994, van Schaik 1983, 1989). However, most studies that have investigated the importance of predation on primates have tended to focus on either patterns across different species (Anderson 1986, Cheney and Wrangham 1987, Goodman *et al.* 1993, Hill and Lee 1998, Isbell 1994) or patterns across individuals within a given population (e.g., Cowlishaw, 1997a, Isbell and Young 1993, Stanford 1995). In contrast, little attention has been paid to the consistency of antipredator strategies across different populations of the same primate species. Nevertheless, such patterns can potentially shed valuable light on intraspecific variation in tolerance to predation risk and the strategies used to reduce that risk, together with the differential costs and benefits of such strategies in different populations.

Baboons represent an ideal taxon for such an investigation, since predation risk in this species is relatively well understood (e.g., Cowlishaw 1994, 1997a) and the wealth of data available on their ecological and behavioral flexibility (Barton *et al.* 1996) permits detailed assessments on the impact of predation risk across a wide array of ecological conditions (e.g., Dunbar 1996). The evidence already available for baboons indicates that antipredator behavior is not fixed but that individuals invest differentially with response to the degree of predation risk. For example, baboons that live in small, high-risk, groups are more active in their use of refuges, such as tall trees and cliff faces, than are larger groups. Importantly, this pattern has been reported in more than one population: Amboseli, Kenya (Stacey 1986) and Tsaobis, Namibia (Cowlishaw 1997b). Likewise, similar patterns of avoidance of food-rich but high-risk habitats by baboons has been reported in two different populations: Tsaobis (Cowlishaw 1997a) and De Hoop, South Africa (Hill 1999). These studies thus emphasize not only the flexibility of baboon antipredator behavior, but also the consistency of antipredator responses between different populations.

The purpose of the present study is to conduct a systematic and detailed investigation of the consistency of antipredator behavior across two different baboon populations: De Hoop in South Africa and Tsaobis in Namibia. We focus our analysis on vigilance, since this is an easily recognizable and quantifiable antipredator behavior that can be directly compared across populations. Moreover, it is a highly flexible behavioral response to predators, and we have a good understanding of the different factors that can affect vigilance in primates. These include group size (Isbell and Young 1993, de Ruiter 1986) and composition (Rose and Fedigan 1995, van Schaik and van Noordwijk 1989), habitat visibility (Chapman 1985, Cowlishaw 1998), refuge proximity (Cowlishaw 1998), and distance to nearest neighbors (Cowlishaw 1998, Treves 1998). Although vigilance might also have functional roles that are not related to predation, for example the detection of potential mates and competitors (Cowlishaw 1998, Treves 2000), the current evidence suggests that this is primarily the case only in male baboons (who are therefore not included in this analysis), and that female vigilance is predominantly related to predation risk (Cowlishaw 1998).

Our analysis first examines whether the performance of antipredator vigilance in foraging females is similar in the two populations. Our analyses show that it is not. In fact, foraging female baboons at De Hoop spend about twice as much time vigilant as those at Tsaobis. Assuming that all foraging baboons respond in a similar way to predation risk, and assuming further that all such individuals will strive to maintain a similar level of safety across populations, we hypothesize that this pattern can best be explained by differences in the relative costs and benefits of antipredator vigilance between the two sites. Specifically, these differences might be accounted for by (a) higher fitness costs of vigilance during foraging at Tsaobis, perhaps due to reduced compatibility between foraging and vigilance at this site, and/or (b) higher fitness benefits of vigilance during foraging at De Hoop, perhaps due to higher predation risk at this site. If our hypothesis is correct, then we should find that the differences in vigilance between the two populations disappear once the local differences in costs/benefits have been statistically controlled.

In order to test this hypothesis, our analysis attempts to control for the different costs and benefits of vigilance that might exist between the two populations. First, we evaluate the costs of vigilance. Specifically, we investigate whether vigilance is more costly during foraging at Tsaobis because the Tsaobis baboons spend more of their foraging time feeding rather than traveling between feeding events. Vigilance is likely to be most costly when performed during the feeding component of foraging time, since it interrupts feeding activities and hence reduces food intake rates (Treves 2000); in contrast, vigilance during the traveling component of foraging is not associated with this cost. Second, we investigate the benefits of vigilance. In order to do this we examine four measures of local predation risk that might cause foraging female baboons to adjust their vigilance (see Cowlishaw 1998 for review and explanation). The first measure is the size of the social group, given that females in smaller groups perceive a higher risk of predation. The second is the distance to refuge, given that baboons closer to refuges are typically safer from predators. The third is the visibility of the habitat, given that leopards attack from close cover, thus making low-visibility habitats more dangerous for baboons. The fourth is the proximity of the nearest neighbor, given that individuals will be at less risk of predation if they are in company rather than alone.

Methods

The data presented here are drawn from females across five groups of chacma baboons (*Papio cynocephalus ursinus*): one group from De Hoop Nature Reserve, South Africa, while the remaining four groups are from Tsaobis Leopard Park, Namibia (see Table 12.1).

Study sites
De Hoop Nature Reserve (20° 24′E, 34° 27′S) is a coastal reserve within the Overberg region, Western Cape Province, South Africa. The reserve is characterized by limestone hills on the western regions, with the Potberg Mountains rising in the northeast. The baboons ranged in an area surrounding the De Hoop Vlei, a large landlocked body of brackish water that is also fed by several freshwater springs. Vegetation is dominated by coastal fynbos, a unique and diverse vegetation type comprising four primary vegetation groups: Proteaceae, Ericaceae, Restionaceae and geophytes. De Hoop experiences a Mediterranean climate and receives most of its rainfall during the austral winter. Mean annual rainfall is 428 mm

Table 12.1. *Study group compositions for the two populations*

| Population | Group | Age/sex class | | | | Group size |
		Adult Male	Adult Female	Juvenile	Infants	
De Hoop	VT	4	12	19	7	42
Tsaobis	A	1	8	10	3	22
	B	2	11	10	1	24
	C	4	13	15	4	36
	D	6	15	30	4	55

with a mean annual temperature of 17.0 °C. The reserve ranges in altitude from 0 to 611 m. A more detailed description of the ecology of the reserve is given in Hill (1999).

Tsaobis Leopard Park (15° 45′ E, 22° 23′ S) is located in the semi-desert Pro-Namib region of Namibia. The reserve is characterized by mountains and ravines, fringed by steep rocky foothills and rolling gravel and alluvial plains. The site is bordered by the ephemeral Swakop River, which supports patches of riparian woodland dominated by *Faidherbia albida*, *Prosopis glandulosa* and *Salvadora persica*. Beyond the woodland, vegetation is sparse with dwarf trees and low shrubs dominated by *Commiphora virgata*. The climate is extremely arid, with a mean annual rainfall of 85 mm and mean annual temperatures of 24.8 °C. Altitude ranges from 683 to 1445 m. A full description of the site is given in Cowlishaw and Davies (1997).

Data collection and analysis

Data for the De Hoop baboons were collected by means of instantaneous point samples (Altmann 1974) of all visible individuals at 30-minute intervals during full day follows, while those for Tsaobis result from instantaneous point samples (Altmann 1974) of a focal individual at 5-minute intervals over multiple one-hour periods per focal. All adult males and females were sampled for De Hoop, with all adult males and selected adult females sampled at Tsaobis. The De Hoop data were collected over a period of 10 months (Mar–Dec 1997); in contrast the Tsaobis data were collected over an intensive 2-month period (Sept–Nov 1991). More detailed descriptions of the study methods at each site are given in Hill (1999) and Cowlishaw (1998). It is unlikely that the differences in methodology between the two studies will lead to significant biases in the data collected, and thus

Table 12.2. *Classification of behavioral and habitat variables incorporated in the analyses*

Behavioral category	Summary of categories
Activity	Foraging (defined as feeding and moving activity combined) or Nonforaging
Vigilance	Yes or No
Habitat type	Closed (visibility <30 m) or Open (visibility >30 m)
Refuge distance	Close (0–20m) or Distant (>20 m)
Neighbor distance	Aggregated (≥1 individual within 5 m) or Dispersed (no individuals within 5 m)

for the remainder of this chapter the data for the two populations are considered wholly comparable. However, the difference in sampling effort between the two populations does appear to result in the Tsaobis females exhibiting a much greater range of values than the De Hoop females. The relationships presented may thus be tighter for the De Hoop females than for the Tsaobis females because they were sampled more extensively, although the broader range of social and ecological conditions at Tsaobis is also likely to contribute to this pattern. At each sample point, the vigilance state of the individual was recorded (an animal was considered vigilant when its head was up and its eyes open), together with its current activity and three measures of its present state of risk: its surrounding habitat type, its distance to refuge and its proximity to neighbors (Table 12.2). In the case of current activity, an animal was considered to be foraging when its activity consisted of either feeding or traveling. Although some traveling time may be unrelated to foraging, for example on the route between the last feeding event of the day and the sleeping site, the great majority of traveling is time spent reaching the next food item or patch. In the case of habitat type, the habitats were divided into two categories on the basis of visibility: closed habitats, where mean visibility at baboon-eye level was below 30 m, and open habitats, where mean visibility was at least 30 m. Three of six habitat types at De Hoop (see Hill 1999), and one of four at Tsaobis (see Cowlishaw 1997a), were thus classified as low visibility. In the case of refuges (trees or cliffs of at least 8 m height and inclined at an angle of at least 75° to the horizontal: Cowlishaw 1997b), two conditions are recognized: individuals in close proximity are defined as those within 20 m of a refuge (including those on a refuge), whereas those distant are at least 20 m away from the nearest refuge. Finally, in the case of neighbor proximity, two categories are once again defined: aggregated, where at least one adult individual is within 5 m of the focal animal, and dispersed, where no adult baboons are within 5 m of the focal individual. Although previous studies have shown primates to be sensitive to the number of near neighbors they have (Treves 1998), it is likely that isolation is the key effect and several previous studies have reported a simple decline in vigilance in the presence of near neighbors (Robinson 1981, van Schaik and van Noordwijk 1989, Steenbeck *et al.* 1999).

If an individual was poorly sampled during foraging, such that five or fewer instantaneous scans were recorded, then this individual was excluded from the analysis (see Cowlishaw 1998, Isbell and Young 1993). All of the proportional data are angular transformed (Sokal and Rohlf 1981) to ensure normality for parametric analysis. However, all figures are drawn utilizing the untransformed data for ease of interpretation (while these figures may only approximate the statistical relationships described in the text, they differ only very slightly from those produced with transformed data). To control for the multivariate nature of vigilance in bivariate plots, the graphs presented typically plot the residual foraging time spent vigilant from the preceding analysis rather than the absolute values, although population identity is not included as a factor in the computation of the residual values.

Results

General patterns of vigilance during foraging
Strong differences exist in the vigilance levels of foraging females between the De Hoop and Tsaobis baboon populations (Fig. 12.1: t-test, $t = 5.90$, df $= 42.0$, $p < 0.001$). Males, who are included here for comparative purposes, also show the same pattern ($t = 4.66$, df $= 19$, $p < 0.001$). In both sexes, the De Hoop baboons spend approximately twice as much time vigilant as the Tsaobis baboons.

Costs of vigilance during foraging
An analysis of the costs of vigilance during foraging indicates that vigilance appears to be incompatible with feeding, since females that spend more of their foraging time feeding are also less vigilant (Fig. 12.2: $r^2 = 0.61$, $F_{(1,46)} = 70.7$, $p < 0.001$). This cost of vigilance

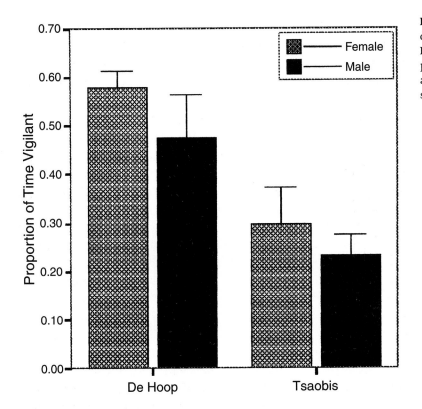

Fig. 12.1. Vigilance rates during foraging for the De Hoop and Tsaobis populations. Mean values and standard errors are shown.

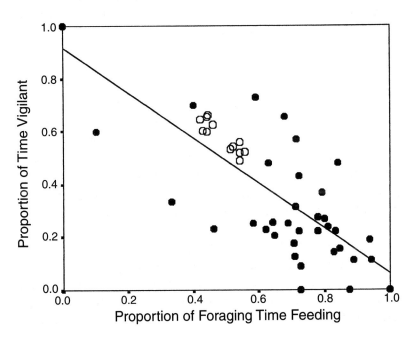

Fig. 12.2. Relationship between proportion of foraging time spent vigilant and proportion of foraging time spent feeding (De Hoop females, open circles; Tsaobis females, solid circles). The line is a best fit least-squares regression for the entire data set.

Fig. 12.3. Relationship between residual proportion of foraging time spent vigilant (calculated from Fig. 12.2) and female group size (De Hoop females, open circles; Tsaobis females, solid circles). The line is a best fit least-squares regression for the entire data set.

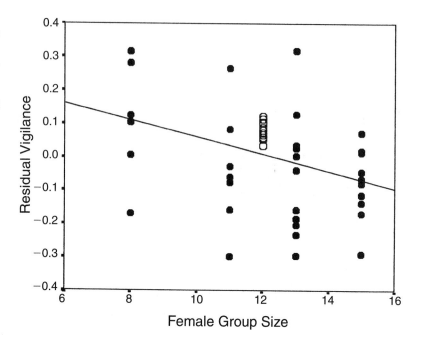

appears to be particularly heavy at Tsaobis, since these baboons spend considerably more of their foraging time feeding compared to De Hoop (De Hoop: 0.49 ± 0.05, Tsaobis: 0.69 ± 0.21, t-test: $t = -4.60$, df$= 41.5$, $p < 0.001$). However, the fact that all the De Hoop females lie above the regression line indicates that these costs alone are not sufficient to explain the relatively high vigilance rates at De Hoop, such that there remains a significant difference between the two populations (analysis of covariance (ANCOVA): $r^2 = 0.64$, $F_{(2,48)} = 40.0$, $p < 0.001$; feeding covariate: $F_{(1,48)} = 53.0$, $p < 0.001$; population factor: $F_{(1,48)} = 4.27$, $p < 0.05$).

Benefits of vigilance during foraging
Effects of group size
Vigilance should become increasingly beneficial to females in smaller groups (indexed here as the number of adult females) due to the greater risk of predation in such groups. Investigation of this pattern (Fig. 12.3) reveals that as predicted, baboons in smaller groups do spend more of their foraging time vigilant. However, there is no apparent difference in average group size between the De Hoop and Tsaobis populations, indicating that this cannot explain the differences that still remain in vigilance between the populations. Indeed, the De Hoop females still exhibit levels of vig-

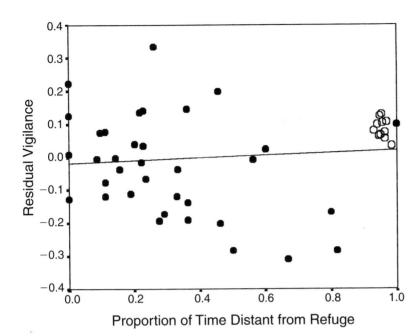

Fig. 12.4. Relationship between residual proportion of foraging time spent vigilant (calculated from Fig. 12.3) and proportion of time spent distant from refuge (De Hoop females, open circles; Tsaobis females, solid circles). The line is a best fit least-squares regression for the entire data set.

ilance that consistently exceed the trend (ANCOVA: $r^2 = 0.69$, $F_{(3,48)} = 32.6$, $p < 0.001$; feeding covariate: $F_{(1,48)} = 46.6$, $p < 0.001$; group size covariate: $F_{(1,48)} = 7.07$, $p < 0.02$; population factor: $F_{(1,48)} = 5.10$, $p < 0.03$). These results do not differ substantially if total group size is used instead of the number of adult females.

Effects of refuge proximity

Vigilance should also become increasingly beneficial to females who are distant from refuge. This is precisely the relationship observed (Fig. 12.4), although the effect is slight. Nevertheless, baboons that are more distant from refuge spend a greater proportion of their foraging time vigilant. Notably, the De Hoop females spend about three times as much foraging time distant from refuge than do the Tsaobis baboons (De Hoop: 0.95 ± 0.01, Tsaobis: 0.30 ± 0.25, t-test: $t = -4.60$, df $= 41.5$, $p < 0.001$). However, once again, the effects of this additional measure of predation risk are not suffi-cient to account for the remaining population differences, with the De Hoop females still exhibiting vigilance levels in excess of that predicted (ANCOVA: $r^2 = 0.75$, $F_{(4,48)} = 31.5$, $p < 0.001$; feeding covari-ate: $F_{(1,48)} = 47.0$, $p < 0.001$; group size covariate: $F_{(1,48)} = 10.8$, $p < 0.005$; refuge distance covariate: $F_{(1,48)} = 9.44$, $p < 0.005$; population factor: $F_{(1,48)} = 15.4$, $p < 0.001$).

Effects of habitat type

Baboons that spend more time in closed habitats should be more vigilant than others, due to the danger of leopard attack in such habitats. However, evaluation of this predicted relationship suggests that it is absent across these two baboon populations, and it is clearly insufficient to explain the persistent differences that remain between the two populations (ANCOVA: $r^2 = 0.75$, $F_{(5,48)} = 24.8$, $p < 0.001$; feeding covariate: $F_{(1,48)} = 46.3$, $p < 0.001$; group size covariate: $F_{(1,48)} = 10.7$, $p = 0.002$; refuge distance covariate: $F_{(1,48)} = 7.25$, $p = 0.01$; habitat covariate: $F_{(1,48)} = 0.22$, $p > 0.60$; population factor: $F_{(1,48)} = 15.4$, $p < 0.001$). This latter point though is perhaps not surprising given the fact that there is no difference in the amount of time that the two populations spend in high-risk habitats (De Hoop: 0.40 ± 0.04, Tsaobis: 0.49 ± 0.32, t-test: $t = -1.38$, df $= 36.6$, $p > 0.10$). Nevertheless, this cannot account for the lack of significance of habitat visibility as a factor determining baboon vigilance levels.

Effects of nearest neighbor proximity

Finally, baboons that spend more time distant from their neighbors should spend more time vigilant because of the higher risk of predation that such dispersed spacing behavior entails. A plot of the proportion of foraging time spent vigilant against the proportion of time spent dispersed (Fig. 12.5) suggests that this is indeed the case. Notably, the De Hoop baboons spend three times as much foraging time in such dispersed positions (De Hoop: 0.75 ± 0.05, Tsaobis: 0.21 ± 0.15, t-test: $t = 14.61$, df $= 46.3$, $p < 0.001$). Crucially, the inclusion of this final component of predation risk appears to be sufficient to account for any remaining differences in the proportion of foraging time spent vigilant between females in the two populations (ANCOVA: $r^2 = 0.77$, $F_{(5,48)} = 28.6$, $p < 0.001$; feeding covariate: $F_{(1,48)} = 49.2$, $p < 0.001$; group size covariate: $F_{(1,48)} = 15.0$, $p < 0.001$; refuge distance covariate: $F_{(1,48)} = 8.50$, $p < 0.01$; neighbor proximity covariate: $F_{(1,48)} = 5.05$, $p = 0.03$; population factor: $F_{(1,48)} = 2.80$, $p > 0.10$).

Discussion

Our results indicate that female baboons at De Hoop spend about twice as much time vigilant when foraging than females at Tsaobis, and that these differences may be explained by differences in the costs and benefits of antipredator vigilance between the two populations. In total, four of the five different measures of the costs and

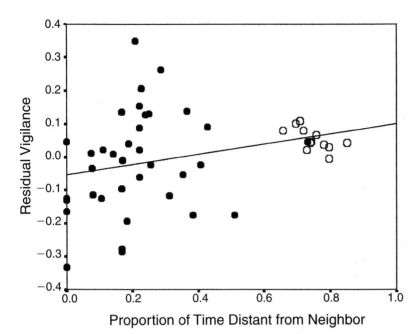

Fig. 12.5. Relationship between residual proportion of foraging time spent vigilant (calculated from Fig. 12.4) and proportion of time spent distant from nearest adult neighbor (De Hoop females, open circles; Tsaobis females, solid circles). The line is a best fit least-squares regression for the entire data set.

benefits tested in this analysis were found to independently contribute to the overall variance in vigilance among foraging females: these were the size of the social group and the proportion of foraging time that was spent feeding, spent distant from refuges, and spent distant from neighbors. The most important of these four measures appears to be the proportion of foraging time spent feeding, which alone accounts for 61% of the variance. Together, the four variables explain over 75% of the variance in the proportion of foraging time spent vigilant, and account for the observed differences between the two populations.

One complication in the interpretation of these results relates to the finding that vigilance during foraging becomes increasingly costly as the proportion of foraging time spent feeding increases. There is an alternative explanation for this pattern: that vigilance becomes increasingly beneficial as the proportion of feeding time declines. In fact, this is exactly what theory would predict, because as feeding time declines during foraging so traveling time increases, and traveling is associated with the highest risk of predation of all baboon activities (Cowlishaw 1998). Unfortunately, it is not possible to distinguish between these two explanations in this analysis; indeed, both might apply equally. Fortunately, though, these explanations are effectively two alternative sides of the same

coin; the only key difference between them is their relative emphasis on costs versus benefits.

During foraging, the De Hoop females appear to be at greater
risk of predation than the Tsaobis females for two key reasons. First,
the De Hoop baboons spent more time distant from refuges in comparison to the Tsaobis baboons. This almost certainly relates to the
fact that De Hoop is a virtually treeless environment (a characteristic of fynbos vegetation: Campbell *et al.* 1979). In contrast, the
baboons at Tsaobis had access to abundant tree refuges while foraging in the woodland habitat (Cowlishaw 1997a,b). Second, the
baboons at De Hoop spent more time distant from neighbors. This
is most likely the result of differences in the spatial distribution of
food at the two sites. At Tsaobis, foraging was largely confined to the
riverine woodland, where the baboons gathered together in or
under the trees to feed on their seeds, pods, flowers, leaves and
bark. At De Hoop, the baboons were often widely dispersed, particularly when foraging in the grassland or burnt fynbos habitats,
where food was at relatively low density in the herb layer, or where
foraging was confined to dispersed digging sites in the vlei habitat.

No relationship was found between vigilance and time spent in
low-visibility habitats. This is surprising, given the importance of
visibility in determining habitat choice in both populations
(Cowlishaw 1997a, Hill 1999) and the fact that such relationships
have been reported in other studies (Chapman 1985; Underwood
1982). However, an analysis of daily vigilance among the Tsaobis
females similarly found no pattern (Cowlishaw 1998). There are
two possible explanations for this. First, this result might be a consequence of the costs of vigilance while feeding, because the closed
habitats are often those with greatest food availability (Cowlishaw
1997a, Hill 1999). Indeed, there is a positive correlation between
time spent in closed habitats and the proportion of foraging time
spent feeding ($r=0.58$, $n=48$, $p<0.001$). This in itself could be
taken as a means to reduce predation risk, since it would reduce
the amount of time an individual needed to spend in high-risk
environments; moreover, there is evidence from both populations
that the high-risk habitats are not exploited as heavily as would be
anticipated on the basis of food availability (Cowlishaw 1997a, Hill
1999). Second, the measure of vigilance used here is not sufficiently detailed to capture subtle variation in the way vigilance is
performed, and that alternative indices such as glance rates (e.g.,
Alberts 1994) might provide greater insight. It is possible that,
rather than increasing vigilance in closed habitats, baboons

might maintain the same level of vigilance but 'parcel' it in a different fashion. For example, in closed habitats, long vigilance periods followed by prolonged feeding bouts might be an inefficient way of monitoring the environment, since the area that can be scanned is small and the additional information that can be gained by long scans will be minimal. However, the visible area needs to be monitored as constantly as possible, since approaching predators must be detected immediately if a successful attack from such close proximity is to be prevented. As a consequence, frequent glances may be a better strategy in low-visibility habitats, because they would allow regular checks for predators. Such a change in vigilance strategy may not be captured by this analysis, since it does not necessarily require any change in the overall time spent vigilant.

It is also important to note that the relationship described here between group size and vigilance rates was not reported in a previous study of vigilance among the Tsaobis baboons (although there was a trend in the predicted negative direction: Cowlishaw 1998). However, this previous work examined daily patterns of vigilance across all activities, and other studies have reported group size effects to be present for certain activities and not others (Blumstein 1996). Indeed, since vigilance patterns differ markedly between activities in the Tsaobis baboon population (Cowlishaw 1998), variation in activity budgets between groups might serve to weaken patterns of covariation between vigilance and group size, hence leading to the failure to detect a clear relationship when data from several different activities are combined.

Overall the results suggest that the baboons at De Hoop are twice as vigilant as those at Tsaobis when foraging because the Tsaobis baboons spend more foraging time feeding (and/or the De Hoop baboons spend more foraging time traveling: see above) and the De Hoop baboons spend more foraging time distant from refuges and neighbors. The longer distances to both refuges and neighbors at De Hoop suggest that foraging was a more dangerous activity there than at Tsaobis, and that the De Hoop baboons responded appropriately with higher levels of vigilance. Given that the presence of leopards, the primary predators of baboons (Cowlishaw 1994), was only confirmed at Tsaobis and not at De Hoop during these studies, these data suggest that baboons may react more sensitively to the conditions that predispose to leopard predation, rather than the presence of leopards themselves. In fact, this may be the most adaptive strategy for baboons, given the difficulties in reliably assessing the presence/absence of leopards and, particularly, given the relatively rapid changes in presence/absence that can occur (e.g., leopards were reported at De Hoop following this study: Henzi *et al.* 2000).

The findings of this study are based on two key assumptions: (a) that baboons in different populations strive to maintain similar levels of safety, and (b) that baboons in different populations respond in similar ways to predation risk. The results presented here lend considerable support to these assumptions, since there was no difference in vigilance between the two populations once the differences in costs and benefits between them were statistically controlled, and both populations follow the same general trend in their response to predation risk (baboons across both populations increase their vigilance in response to a decline in group size, an increase in distance to refuge, and an increase in distance to nearest neighbor). Nevertheless an element of caution must be exercised in this interpretation. For example, although the higher levels of vigilance at De Hoop might be thought to indicate that foraging activities are twice as dangerous there than at Tsaobis, it might instead be argued that foraging is actually more dangerous at Tsaobis, specifically because the baboons are less vigilant during this activity. Similarly, it is possible that one of the reasons why the De Hoop baboons foraged at greater distances from refuges and neighbors was because they have a higher tolerance to the same level of local predation risk than those at Tsaobis; that is, they are not striving to maintain similar levels of safety (perhaps because leopards are less common residents at De Hoop).

In relation to this latter point, however, it might be possible to make some estimate of the levels of risk that the two populations should be willing to experience. Hill and Dunbar (1998) showed that across species, observed predation rates are closely related to the reproductive rate of that species. Predation rates were thus suggested to represent the rate of loss that animals could accommodate within their life history strategies, such that the animals were adopting behavioral strategies that reduced risk to some undefined 'acceptable' level. Such relationships should hold true within species, with the level of risk a population should be willing to experience being related to potential reproductive rates at that site. Mean interbirth interval at De Hoop is 25 months (Barrett *et al.* 1999). Although similar data are not available for Tsaobis, it is possible to estimate the interbirth interval for this population using the equation given in Hill *et al.* (2000). On the basis of the available ecological and demographic data for Tsaobis, the predicted mean interbirth interval would be 23.0 months. Some caution needs to be employed with this estimate, since the available temperature data for Tsaobis are drawn from Karibib, a locality about 65 km to the north of the study area. Nevertheless, the predicted value is not substantially different from the figure for De Hoop, and given the error margins involved, such differences are unlikely significantly to affect the levels of risk that the two populations perceive as acceptable. As a consequence, it seems likely that risk tolerances of the two populations should be virtually the same. However, such conclusions must be considered preliminary at best, and merely serve to highlight one possible way in which risk tolerances might be assessed. As a consequence, the preceding discussion clearly highlights that our understanding of interpopulation differences in risk tolerance, and the impact this has on the consistency of antipredation strategies, is still somewhat limited. Thus, while our analyses support the notion that baboons (and other primates) are following a species-typical antipredation strategy that is tailored to the local environmental conditions, such conclusions are contingent on our starting assumptions. A primary goal for future research should therefore be to determine whether such assumptions can be justified.

Acknowledgments

We thank Lynne Miller for inviting us to contribute to this volume and to the symposium at the XVIIIth Congress of the International Primatological Society in Adelaide. Attendance at the conference by RAH was aided by a British Academy Conference Grant. RAH thanks Cape Nature Conservation, Peter Henzi and Louise Barrett for permission to work at De Hoop, Tony Weingrill for assistance in the field and Robin Dunbar for financial and logistical support. GC thanks Peter Bruce, Jonathan Davies, Robin Dunbar and Sanjida O'Connell for their invaluable help and support, and August Juchli and the Ministry of Wildlife, Conservation and Tourism for permission to work at Tsaobis Leopard Park and in Namibia, respectively. Finally, we thank Lynne Miller and two anonymous referees for constructive comments on an earlier draft of this manuscript.

REFERENCES

Alberts, S.C. (1994). Vigilance in young baboons: effects of habitat, age, sex and maternal rank on glance rate. *Animal Behaviour* **47**: 749–55.

Alexander, R.D. (1974). The evolution of social behaviour. *Annual Review of Ecological Systems* **5**: 325–83.

Altmann, J. (1974). Observational study of behaviour: sampling methods. *Behaviour* **49**: 227–67.

Anderson, C.M. (1986). Predation and primate evolution. *Primates* **27**: 15–39.

Barrett, L., Henzi, S.P., Weingrill, T., Lycett, J.E., and Hill., R.A. (1999). Market forces predict grooming reciprocity in female baboons. *Proceedings of the Royal Society, London, Series B* **266**: 665–70.

Barton, R.A., Byrne, R.W., and Whiten, A. (1996). Ecology, feeding competition and social structure in baboons. *Behavioural Ecology and Sociobiology* **38**: 321–29.

Blumstein, D.T. (1996). How much does social group size influence golden marmot vigilance? *Behaviour* **133**: 1133–51.

Campbell, B.M., McKenzie, B., and Moll, E.J. (1979). Should there be more tree vegetation in the Mediterranean climate region of South Africa? *Journal of South African Botany* **45**: 453–7.

Chapman, C. (1985). The influence of habitat on behaviour in a group of St. Kitts green monkeys. *Journal of Zoology, London* **206**: 311–20.

Cheney, D.L., and Wrangham, R.W. (1987). Predation. In: B.B. Smuts, D.L. Cheney, R.M. Seyfarth, R.W. Wranham and T.T. Struhsaker, eds., *Primate Societies*, Chicago: University of Chicago Press, pp. 227–39.

Cowlishaw, G.C. (1994). Vulnerability to predation in baboon populations. *Behaviour* **131**: 293–304.

Cowlishaw, G.C. (1997a). Trade-offs between foraging and predation risk determine habitat use in a desert baboon population. *Animal Behaviour* **53**: 667–86.

Cowlishaw, G.C. (1997b). Refuge use and predation risk in a desert baboon population. *Animal Behaviour* **54**: 241–53.

Cowlishaw, G.C. (1998). The role of vigilance in the survival and reproductive strategies of desert baboons. *Behaviour* **135**: 431–52.

Cowlishaw, G.C., and Davies, J.G. (1997). Flora of the Pro-Namib Desert Swakop River catchment, Namibia: community classification and implications for desert vegetation sampling. *Journal of Arid Environs* **36**: 271–90.

Crook, J.H., and Gartlan, J.S. (1966). Evolution of primate societies. *Nature* **210**: 1200–03.

de Ruiter, J.R. (1986). The influence of group size on predator scanning and foraging behaviour of wedge-capped capuchins (*Cebus olivaceus*). *Behaviour* **98**: 240–58.

Dunbar, R.I.M. (1988). *Primate Social Systems*. London: Chapman & Hall.

Dunbar, R.I.M. (1996). Determinants of group size in primates: a general model. In: W.G. Runciman, J. Maynard Smith and R.I.M. Dunbar, eds., *Evolution of Social Behaviour Patterns in Primates and Man*, Oxford: Oxford University Press, pp. 33–57.

Goodman, S.M., O'Connor, S., and Langrand, O. (1993). A review of predation on lemurs: implications for the evolution of social behaviour in small, nocturnal primates. In: P.M. Kappeler and J.U. Ganzhorn, eds., *Lemur Social Systems and their Ecological Basis*, New York: Plenum Press, pp. 51–66.

Henzi, S.P., Barrett, L., Weingrill, T., Dixon, P., and Hill, R.A. (2000). Ruths amid the alien corn: males and the translocation of female chacma baboons. *South African Journal of Science* **96**: 61–2.

Hill, R.A. (1999). Ecological and Demographic Determinants of Time Budgets in Baboons: Implications for Cross-Populational Models of Baboon Socioecology. Ph.D. Thesis, University of Liverpool.

Hill, R.A., and Dunbar, R.I.M. (1998). An evaluation of the roles of predation risk and predation rate as selective pressures on primate grouping behaviour. *Behaviour* **135**: 411–30.

Hill, R.A., and Lee, P.C. (1998). Predation risk as an influence on group size in cercopithecoid primates. *Journal of Zoology, London* **245**: 447–56.

Hill, R.A., Lycett, J.E., and Dunbar, R.I.M. (2000). Ecological and social determinants of birth intervals in baboons. *Behavioural Ecology* **11**: 560–4.

Isbell, L.A. (1994). Predation on primates: ecological patterns and evolutionary consequences. *Evolutionary Anthropology* **3**: 61–71.

Isbell, L.A., and Young C.P. (1993). Social and ecological influences on activity budgets of vervet monkeys, and their implications for group living. *Behavioural Ecology and Sociobiology* **32**: 377–85.

Robinson, J.G. (1981). Spatial structure in foraging groups of wedge-capped capuchin monkeys *Cebus nigrivittatus*. *Animal Behaviour* **29**: 1036–56.

Rose, L.M., and Fedigan, L.M. (1995). Vigilance in white-faced capuchins, *Cebus capucinus*, in Costa Rica. *Animal Behaviour* **49**: 63–70.

Sokal, R.R., and Rohlf, F.J. (1981). *Biometry*. San Francisco: Freeman.

Stacey, P.B. (1986). Group size and foraging efficiency in yellow baboons. *Behavioural Ecology and Sociobiology* **18**: 175–87.

Stanford, C.B. (1995). The influence of chimpanzee predation on group size and antipredator behaviour in red colobus monkeys. *Animal Behaviour* **49**: 577–87.

Steenbeck, R., Piek, R.C., van Buul, M., and van Hooff, J.A.R.A.M. (1999). Vigilance in wild Thomas's langur (*Presbytis thomasi*): the importance of infanticide risk. *Behavioural Ecology and Sociobiology* **45**: 137–50.

Treves, A. (1998). The influence of group size and neighbours on vigilance in two species of arboreal monkeys. *Behaviour* **135**: 453–81.

Treves, A. (2000). Theory and method in studies of vigilance and aggregation. *Animal Behaviour* **60**: 711–22.

van Schaik, C.P. (1983). Why are diurnal primates living in groups? *Behaviour* **87**: 120–44.

van Schaik, C.P. (1989). The ecology of social relationships among female primates. In: V. Standen and R.A. Foley, eds., *Comparative Socioecology: The Behavioural Ecology of Humans and Other Mammals*, Oxford: Blackwell Scientific Publications, pp. 195–218.

van Schaik, C.P., and van Noordwijk, M.A. (1989). The special role of male

Cebus monkeys in predation avoidance and its effects on group composition. *Behavioural Ecology and Sociobiology* **24**: 265–72.

Underwood, R. (1982). Vigilance behaviour in grazing African antelopes. *Behaviour* **79**: 81–107.

13 • Foraging and safety in adult female blue monkeys in the Kakamega Forest, Kenya

MARINA CORDS

Introduction

Finding food and avoiding being someone else's food are two problems every primate must solve. The trick is finding a solution to each problem which does not simultaneously compound the other one. For example, being in close proximity to conspecifics, who can reduce the chances of being preyed upon in various ways, may also lead to increased competition for resources. Feeding on the newly emerged leaves of a deciduous tree, a relatively easily harvested source of protein, may mean forsaking protective cover. Understanding how animals balance the sometimes conflicting demands of efficiency and safety in food acquisition can help us clarify their biological priorities. Comparisons of the various solutions across taxa can help us determine the extent to which behavioral solutions are phylogenetically canalized, or flexible responses to local environmental conditions.

This chapter considers the relations between foraging and antipredator strategies in an African forest guenon, the blue monkey (*Cercopithecus mitis stuhlmanni*). Blue monkeys are omnivores whose major dietary constituents are fruits, leafy matter (including leaf blades, petioles and buds), and invertebrates. They also eat flowers, nectar, gum, seeds, galls, and fungi. They harvest their foods from a broad array of plant species. For example, Cords (1986) reported that blue monkeys at Kakamega used at least 104 plant species as sources of plant food over a 12-month period, and at least 80 plant species (many of which did not double as sources of plant parts) as sources of invertebrates. In addition, blue monkeys show considerable variation in diet over various spatial and temporal scales. Lawes (1991) has described dietary variation that occurs over a broad geographical scale, while Rudran (1978) and Butynski (1990) have compared blue monkey groups that inhabit different parts of

the same study site. Many studies have documented substantial seasonal variation in diet (Beeson 1989, Cords 1986, Kaplin *et al.* 1998,
Lawes *et al.* 1990, Rudran 1978): in general, blue monkeys appear to
feed mainly on fruits when they are available, and to eat more
leaves when fruits are scarce. The contribution of invertebrates to
the diet varies considerably (from roughly 5% to 45% of feeding
records, Lawes 1991). The dietary flexibility that blue monkeys are
evidently able to tolerate would seem to enable them to negotiate
various feeding strategies. I focus here on how these animals adjust
their feeding behavior in response to predation risk.

An important predator on blue monkeys in equatorial forests,
including the site of my study, is the African crowned eagle,
Stephanoaetus coronatus (Cords 1987, Struhsaker and Leakey 1990). It
is an ambush predator that swoops into the monkey group, apparently relying on speed and surprise to capture prey (Brown 1971,
Brown *et al.* 1982, Cords 1987, Zimmerman *et al.* 1996). Monkeys
respond by alarm calling and diving into dense foliage, where they
remain for up to 60 minutes (Cordeiro 1992, Cords 1987). Blue
monkeys also make alarm calls to a variety of small wild mammals
(including mongooses, palm civets, African civets, and various
felids), snakes, and domestic dogs and people (Cords 1987).
However, I will focus on the strategies that they use in avoiding predation by crowned eagles, because responses to this predator are
the most frequent and most extreme, and the components of predation risk (Lima and Dill 1990) that apply to eagles are best understood. In particular, blue monkeys seem to rely primarily on early
detection of the predator which allows them to take cover quickly.
Taking cover may be both a way of avoiding detection by the predator, and a way of thwarting its attack, since large eagles cannot
penetrate the dense foliage where blue monkeys hide. Early detection results from each individual's own vigilance, but also from the
warnings of nearby group-mates. Indeed, Treves (1999) has shown
that individuals' vigilance rates are lower when their group-mates
are in close proximity, indicating that vigilance is shared with near
neighbors. Taking cover is a matter of finding dense clumps of
foliage.

In view of the importance of early warnings from nearby group-
mates and vegetative cover as protection mechanisms, this report
focuses on how blue monkey adult females deploy their time in
feeding trees that vary in terms of what they offer as food, how protected the feeding animal would be by the tree's foliage, how much
space there is for other animals to be feeding in the same tree, and

how many animals actually do feed in the same tree. If blue monkeys simply minimized risk of predation, one might expect them to feed most often in trees with dense foliage, and those that are shared with other (vigilant) conspecifics. First, I present descriptive data to assess the degree to which these expectations are met, both overall and as a function of the type of food being eaten. Then, I summarize data from previous work relating to the importance of feeding competition. Next, I evaluate two hypotheses related to trade-offs that blue monkeys may need to make in accommodating their needs for safety and nutrition. The first is that the monkeys forsake the benefit of vigilance from nearby conspecifics more when feeding on foods for which competition is most keen. The distribution and relative abundance of primate foods have been recognized as predictors of the degree of competition associated with food consumption (Isbell 1991, Isbell *et al.* 1998, van Schaik 1989). Fruit, because it is clumped and monopolizeable, and insects, because they are rare, have generally been identified as foods over which primates compete more, whereas leaves, being generally both less clumped and more abundant, seem to be less associated with competition. The competition that occurs over fruit and insects is not likely to be identical: fruit feeding sites can be usurped through contest competition, whereas competition over invertebrates is more likely to be exploitative. In both cases, however, one might expect individuals to minimize competition by spreading out. The second hypothesis is that the monkeys trade off the benefits of foraging in dense foliage and foraging near conspecifics; this would be expected if these are alternative strategies for minimizing predation risk. Finally, in discussing the results, I include some consideration of the role that mixed-species associations may play in the antipredator strategies of blue monkeys, even though these associations were not part of the data set analyzed in this chapter.

Subjects and methods

Data were collected from the habituated Tw group of blue monkeys in the Kakamega Forest, Kenya (see Cords 1987 for a description of this rain forest site). This group has been studied intermittently since 1979. The data reported here were collected during two 3-month (June–August) periods in 1992 and 1993. These months represent a relatively wet time of year (Cords 1987), in which there are typically few fruiting trees available. During this period, the group varied in size from 31 to 34, excluding adult males. On most days there was just one adult male in the group, although occasional visits from other males did occur as part of the mating season, which ran concurrently with this study.

Crowned eagles do occur at Kakamega, where they are heard or seen approximately every 2–3 days. No successful attack on blue monkeys has been witnessed in over 20 years' research; however, the monkeys' very strong response to crowned eagles, to the eagle alarm calls of neighboring groups of primates, and to eagle-like stimuli (e.g., any large soaring bird) suggests that these birds pose a real threat. Dogs and people are the only predators known to have killed blue monkeys in the last 20 years at Kakamega, but many other disappearances of study animals are likely to have been caused by natural predators, including eagles.

Focal animal samples were collected on 16 females, including all adult (parous) females in the study group, as well as those nulliparous females who had attained adult body size, and whose behavior generally resembled that of adults rather than juveniles. In particular, these females were not observed to play, and they engaged in extensive mutual grooming with several adult partners, whereas younger juveniles focus grooming more on their own mothers (Cords 2000a). Female blue monkeys first give birth at about 6–7 years of age, and all nulliparous subjects gave birth for the first time within 2 years of being sampled. There were 16 subjects in all, including 12 parous and three nulliparous females in 1992, and 14 parous and one nulliparous female in 1993. For the 14 subjects sampled in both years, data were combined across years. Conditions were comparable in the 2 years, in that rainfall patterns were similar, and the social situation of the group was stable (i.e., there was no multi-male influx during the mating season).

Focal samples were not collected on adult males or juveniles. Because of their vigilance for rivals, especially during the mating season, adult male blue monkeys seem to operate under a different set of constraints than females, and should be considered separately. With only one adult male regularly available to sample, however, I could not achieve a satisfactory sample of male behavior. Juveniles were not included because they could not be reliably identified as individuals.

The scheduling and procedural rules of the focal sampling are described by Cords (2000b), and briefly recapitulated here. I conducted focal animal samples totaling 5 h per female in 1992 and 11 h per female in 1993. I followed each subject's activities using binoculars and a grid of paths, but I left paths when necessary to maintain visual contact. Each focal sample was scheduled to last 1 h; however, if the subject disappeared from view for 15 minutes, the sample was terminated at the moment of disappearance, and resumed at the next opportunity, usually on the same day. If the subject disappeared for less than 15 minutes, the sample was extended until she had been in view for a total of 60 minutes. A monkey usually disappeared because she moved into dense vegetation, not because she was engaged in behavior that was difficult to follow. The proportion of time spent in dense vegetation may therefore be somewhat under-represented in the analysis, but this under-representation should occur independent of behavioral category at the time of disappearance.

A predetermined sampling schedule was impractical, given the observation conditions and the habit of blue monkeys at Kakamega to disperse over several hundred meters. Instead, I spread observations of each individual across the two sampling periods, across the hours of the day, and across the different forest types that the group occupied (see Cords 2000b for details).

During focal samples, I continuously recorded the subject's activities and noted transition times (to the nearest second) between activity classes. Whenever the subject moved from one plant to another, I also noted the time and the plant's identity. 'Feeding' included ingestion, brief movements between ingestion sites, and invertebrate foraging by manual searching and visual searching (when the subject obviously gazed at nearby substrates). I also noted transition times between different plant parts consumed. Most of the analyses focus on time spent feeding, which averaged 56% of observation time (SD = 8%, range 42%–67%) across the 16 subjects. Other activities that I monitored included 'resting,' 'moving' and 'socializing' (Cords 2000b).

When the subject fed in a tree (or liana), the number of other individuals in the same tree was continuously monitored to assess the potential for feeding competition. Where an exact count was impossible, I noted whether the tree contained some (≥1) or many (≥5) other monkeys. Cords (2000b) describes the way in which individual trees, lianas and liana tangles were distinguished.

Feeding trees were further characterized by size, which was expressed in terms of how many feeding monkeys the tree could hold. This metric was adopted because it incorporated the monkeys' perspective better than arbitrary measurements such as crown diameter. For example, trees with deep crowns, or many branches, might have many potential feeding sites, despite a small diameter. On the basis of 15 years' experience of how blue monkeys at this site typically space themselves within different trees, three size classes were distinguished: 'small' trees could be occupied by no more than two animals, and typically had crown diameters ≤2 m. 'Medium' trees (with typical crown diameter of 2–8 m) could be occupied by no more than three to eight individuals. 'Large' trees (with typical crown diameter ≥8 m) could be occupied by a maximum number of nine or more monkeys. In the analysis, I distinguished trees that were full (at least one other monkey in a small tree, at least three in a medium-sized tree, and at least six or 'many' in a large tree) from those that were not full. Large trees were the hardest to evaluate in terms of how full they were (since the largest crowns were the least likely to be visible in their entirety, and/or most likely to accommodate a large number of monkeys that was difficult to count quickly), and a cut-off below the nine or more monkeys that such trees could actually hold was chosen for this reason; a higher cut-off would have resulted in almost no data on the fullness of large trees. I have noted where this adjustment may have influenced results.

Data on the density of foliage surrounding feeding monkeys were not collected concurrently with behavioral observations. Instead, foliage density (high, medium, and low) was assigned after the fact according to which species of tree the monkey occupied. The assignment of typical foliage densities to each tree species was based on my knowledge of the trees at this site, gained over 15 years of field work (Cords 1995). A monkey was very conspicuous in trees with low foliage densities, where most or all of its body was exposed; in trees with high foliage densities, only a part of the animal was visible, and usually from limited, relatively nearby vantage points (Cords 1990a). The validity of these measures of foliage density as reflective of perceived risk have been previously explored in the same study population: I found that individuals showed more vigilance behavior (looking up) as foliage density decreased (Cords 1990a).

Results

The diet of the blue monkeys in this study differed in some ways from earlier reports. The 16 subjects used a total of at least 83 plant species as sources of plant and invertebrate foods (combined). They averaged 46% (SD 12%, range 16%–62%) of feeding time on foliage,

37% (SD 12%, range 21%–62%) on fruits, and 12% (SD 7%, range 4%–24%) on invertebrates, the three major components of the diet. These figures indicate considerably more concentration on foliage, and a lower consumption of fruit and invertebrates, than Cords (1986) previously reported for adult female blue monkeys in the same population (23% foliage, 49% fruit, 18% invertebrates). These differences are likely to reflect mainly the fact that the current observations occurred during a seasonal period of fruit scarcity, and perhaps also the difference in methods used to score feeding behavior. The earlier report was based on a particular type of feeding frequency score (rather than measures of duration), and such scoring often emphasizes items (like insects and some fruits) that are quickly grabbed and swallowed. Regardless of the differences, however, it is clear that even during the period when the present observations were made, blue monkeys maintain a diverse diet. This should allow them flexibility in their feeding-plus-safety strategies.

Most feeding ($\bar{x}=78\%$, SD 11%, range 72%–85%, $n=16$ females) occurred in trees with intermediate foliage densities. Only an average of 5% (SD 0.7%, range 2.5%–5.5%) of feeding time occurred in trees with low foliage densities, and 18% (SD 10%, range 11%–20%) occurred in trees with high foliage densities. Other activities were more likely than feeding to occur at low foliage densities: on average, 15% of moving and 12% of resting took place in trees with low foliage densities (Friedman Test, $F_R=24.38$, 3 df, $n=16$ females, $p=0.0001$, with multiple comparisons, $\alpha=0.05$; the proportion of social time in trees with low foliage density, which averaged 11%, was not significantly different from the proportion of feeding time in such trees).

On average, over two thirds of feeding on each of the three major dietary items occurred in trees with intermediate foliage densities (Table 13.1). Invertebrate feeding, however, was on average two to five times more likely to occur at low foliage densities than was fruit feeding or leaf feeding, which did not differ significantly from each other (Friedman Test, $F_R=8.79$, 2 df, $n=16$ females, $p=0.0124$, with multiple comparisons, $\alpha=0.05$).

Most feeding ($\bar{x}=58\%$, SD 13%, range 29%–76%, $n=16$ females) occurred in large trees. Medium-sized trees were occupied for an average of 25% (SD 11%, range 11%–44%) of feeding time, and small trees an average of 18% (SD 7%, range 4%–26%) of feeding time. Fruit feeding was especially likely to occur in large trees and unlikely to occur in small trees, whereas leaf feeding (especially on mature

Table 13.1. *Average time spent feeding in trees with different foliage densities as a function of the item consumed* (n = 16 *females*)

Foliage density	Percentage of invertebrate feeding	Percentage of fruit feeding	Percentage of leaf feeding
High	21 (1.3)	9 (1.5)	9 (1.2)
Medium	68 (7.7)	86 (35.2)	89 (42.0)
Low	11 (3.0)	5 (3.9)	2 (4.2)

Notes:
The number in parentheses represents the average percentage of total feeding time devoted to a particular item at a particular foliage density.

Table 13.2. *Average time spent feeding in trees of different size as a function of the item consumed* (n = 16 *females*)

Tree size	Percentage of invertebrate feeding	Percentage of fruit feeding	Percentage of leaf feeding
Small	12 (1.6)	6 (2.4)	28 (13.8)
Medium	29 (3.7)	15 (4.7)	34 (16.1)
Big	59 (6.9)	80 (32.3)	38 (18.6)

Notes:
The number in parentheses represents the average percentage of total feeding time devoted to a particular item at a particular tree size.

leaves) was least likely to occur in large trees and most likely to occur in small trees (Table 13.2; Friedman Test, $F_R = 19.63$, 2 df, $n = 16$ females, $p = 0.0001$). For a feeding monkey, invertebrate and fruit feeding are probably more visually demanding than leaf feeding: close attention to mobile prey or the matrix from which they must be extracted and selection of the ripest fruits require a high level of visual attention. In this context, it is noteworthy that when blue monkeys feed on invertebrates and fruits, they do so in trees where they could be accompanied by other group members.

Surprisingly, however, these blue monkey females spent a lot of time feeding in trees by themselves (Fig. 13.1). An average blue monkey female had only 1.8 other monkeys in the same tree, and

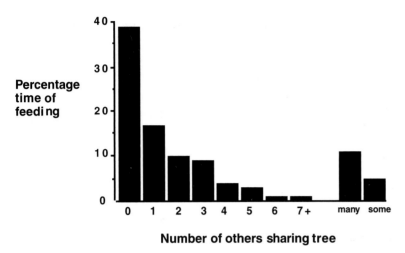

Fig. 13.1. The percentage of feeding time that an average female spends with different numbers of group-mates in her feeding tree. 'Many' refers to five or more, while 'some' refers to one or more. To calculate the average number of monkeys in a feeding tree, observations contributing to these two imperfectly specified categories were added to columns 5–7 (for 'many') and columns 2–7 (for 'some'), distributed in proportion to the percentages originally in those columns.

spent 39% (SD = 9.3%, range 25%–64%) of her feeding time in a tree by herself. An average female spent 76% (SD 13%, range 48%–95%, $n = 16$) of her feeding time in trees that were not 'full,' and so could have accommodated additional monkeys. Larger trees were somewhat more likely to be full than were medium-sized or small trees: an average of 29% of feeding in large trees occurred when they were fully occupied, whereas the corresponding percentages for medium-sized and small trees were each 17%. The higher figure for large trees may, however, reflect the less stringent criteria for assessing fullness in these crowns (see p. 210). In any case, it appears that female blue monkeys at Kakamega spread out to feed in a less dense spatial distribution than would be physically necessary.

Spreading out may be a way to minimize competition for food. Indeed, there is evidence from this population that feeding with others is associated with aggressive interference competition. Cords (2000b), analyzing the same focal animal samples discussed here, reported that 93% of agonism involving adult females occurred in feeding (and drinking) trees, even though the average percentage of time spent feeding was only 54%. Furthermore, these contests did not occur proportionately to the time spent feeding on various items: agonism while feeding on fruit occurred 1.7 times more often than expected based on the time spent feeding on fruit. Agonism while feeding on invertebrates occurred 0.3 times as often as expected, while agonism during foliage feeding occurred 0.75 times as often as expected. The loser of agonistic contests usually moved to a new feeding site in the tree, or left it entirely.

If spreading out in this way is a strategy to minimize competition for food, then one might predict that monkeys should spread out, especially when feeding on those foods which engender most competition. I analyzed the relationship between dietary items and the number of group-mates in the tree separately for large and for medium-sized trees, since the number of other monkeys in a tree will be affected by how much space is available to accommodate them. (Small trees did not allow for much variation in the number of other individuals and were therefore not analyzed.) In both medium-sized and large trees, feeding alone was about twice as likely when the animals were feeding on invertebrates than when they fed on fruits or leaves. In large trees, the average female spent 46% (SE=5%, range 11%–94%, n=16) of her invertebrate feeding time with no other monkeys in the tree, but only 13% (SE 4%, range 0%–57%) of her fruit-feeding time and only 17% (SE 7%, range 0%–100%) of her leaf-feeding time alone in the tree (Wilcoxon Matched Pairs Signed Ranks test, Z=−3.31, p=0.0009 for fruits, Z=−2.844, p=0.0045 for leaves). In medium-sized trees, the average female spent 86% (SE=3%, range 62%–100%) of her invertebrate feeding time alone in the tree, but only 40% (SE 23%, range 5%–90%) of her fruit-feeding time alone (Wilcoxon test, Z=−2.22, p=0.026), and only 45% (SE 13%, range 0%–100%) of her leaf-feeding time alone (Wilcoxon test, Z=−3.46, p=0.005). In trees of either size, there was no significant difference in females' likelihood of feeding alone on fruits versus leaves. These results show that females are especially likely to spread out when feeding on invertebrates. They also show that females are more likely to feed alone, whatever the item consumed, in smaller trees than in larger trees.

The number of other monkeys in a feeding tree is related not only to what is being eaten, but also to foliage density. If feeding with others and feeding in dense foliage are two ways of increasing an individual's safety, one might expect that animals would be more likely to feed with others when feeding in trees with lower foliage density. To see if such a trade-off existed, I analyzed data on feeding in large, medium-sized and small trees separately, again to control for the opportunities that exist in trees of different size to feed with others. I looked for a monotonic decrease across trees with high, medium and low foliage densities in the percentage of time each female spent alone. For large and small trees, most females increased their time alone as foliage density increased (Table 13.3). In medium-sized trees, this pattern was not apparent.

Table 13.3. *The relationship between foliage density in feeding trees and the tendency of females to feed alone in the tree*

	Small trees	Medium trees	Large trees
Females who fit the pattern[a]	10	8	14
Females who do not fit the pattern	3	7	2
Females who show no difference	3	1	0
One-tailed p-value, Sign Test	0.046	>0.05	0.002

Notes:

[a] For females who 'fit the pattern,' the percentage of time feeding alone was lowest in trees with low foliage density, intermediate in trees with a medium foliage density, and highest in trees with high foliage density.

Discussion

In general, blue monkey adult females seem to feed in trees that provide at least some protection in terms of dense foliage: overall, very little of their feeding occurs in trees with the lowest foliage densities where they would be most conspicuous targets to aerial predators. Other activities, such as resting and moving, were more likely to occur in trees with the lowest foliage densities. Comparisons between these activities and feeding do not, however, allow one to conclude that the monkeys were choosing to minimize their exposure while feeding. Animals can feed only where food is available, and it is also possible that food was not available in trees with low foliage densities. Indeed, only nine of the 83 plant species used for feeding had low foliage densities. More precise measurements of the availability of food as it relates to foliage density would be needed to investigate this possibility further.

The question remains why blue monkeys do not avoid exposed and presumably dangerous areas while moving and resting as thoroughly as they do while feeding. Efficiency of travel routes might be one reason for using open areas for movement. Resting, however, could occur anywhere in principle, so exposed and dangerous places could presumably be avoided even more than they are. One possible explanation is that resting sometimes occurs in open places for thermoregulatory reasons: indeed, blue monkeys at Kakamega regularly sit in exposed, sunny places when ambient temperatures are low, such as in the early morning or when the sun re-emerges after heavy rains.

Feeding monkeys were more likely to forsake the cover of dense foliage when eating invertebrates than when they fed on plant foods. Blue monkeys in equatorial forests harvest insects after prolonged visual and manual searching, often unrolling leaves or prying under bark or through dead woody tissue (Cords 1987, Rudran 1978). These activities require concentrated visual attention to the substrate and the prey. Such visual concentration has been shown to decrease vigilance rates of individual monkeys relative to what they would be when feeding on plant foods (Cords 1990a). Therefore, it is surprising that invertebrate foraging, which is evidently dangerous by its very nature, occurs relatively often in trees that offer little protective cover. This fact may simply reflect the distribution of invertebrate prey, which clearly are available in trees with low foliage densities. Alternatively, perhaps when foraging is especially demanding of visual attention, it becomes important for the monkeys to avoid obstructive cover that would reduce their ability to detect predators quickly, even if it makes them more conspicuous; according to this argument, it might be safer, under certain circumstances, to avoid dense foliage (Treves, Chapter 14 this volume).

Proximity to group mates is another way that blue monkeys can protect themselves from predation, since they can rely on shared or enhanced vigilance of nearby neighbors. Treves (1999), analyzing vigilance rates as a function of the proximity of neighbors in another population of blue monkeys, demonstrated that a feeding individual's time spent scanning was related to the relative numbers of neighbors that are closer versus farther away. The analysis presented here did not directly measure the number of monkeys in close proximity, since the focus was on individual feeding trees, rather than interindividual distances. It is possible that a monkey may have had no neighbors in its tree, but was still close to a conspecific in a neighboring tree. It is also possible that a large crown populated by only a few blue monkeys involved interindividual distances in excess of the 2-m cut-off that Treves identified as critical. However, I suspect that these circumstances did not occur often enough to make the current results misleading: in general, conspecifics in the same tree were within a few meters of one another, and so could presumably have served to reduce predation risk through shared vigilance.

While blue monkeys did share their feeding trees with others most of the time, they also spent a considerable portion (39%) of their time feeding in trees entirely by themselves. The average number of group-mates in the tree was less than two. It appears that this tendency to feed alone or with just a few others is not forced upon the animals by the architecture of the tree crowns. Most of their feeding occurred in large trees, and most of it occurred in trees that could have accommodated additional individuals. Thus it seems that blue monkeys generally choose to feed with fewer group-mates nearby (in the same crown) than would be possible.

There are at least two, not mutually exclusive, explanations for this observation. First, there are costs to feeding with others in close proximity, particularly in terms of interference competition, which may limit its occurrence. Rates of aggression were disproportionately high while the monkeys fed on fruit, which tended to occur in large trees that may have been 'full' more often. The fact that monkeys were more likely to feed with others in larger trees is also consistent with the idea that there was some ideal spacing between individuals that can be achieved more easily in the largest trees. That spacing could be related to feeding efficiency as a function of competition by others in the same crown. Similar conclusions have been drawn by Phillips (1995a,b) from her study of white-faced capuchin monkey (*Cebus capucinus*) feeding. Second, the protective benefits of feeding in close proximity to others may be achieved in alternative ways: there was some evidence, for example, that monkeys traded off the protection of close neighbors for the protection of higher foliage density, at least in the smallest and largest trees.

In the context of having others in the same tree, feeding on invertebrates was again a surprising case. Blue monkey females were twice as likely to be alone in a tree when feeding on invertebrates rather than plant foods, despite the fact that invertebrate feeding depresses individual vigilance levels. Invertebrate feeding seems to be carried out in the most dangerous circumstances – relatively often in exposed tree crowns and without the presence of group-mates. Reasons for using exposed crowns for invertebrate feeding were discussed above. Feeding alone may result from the ability of invertebrates to change their activity or position in response to the foraging activities of a monkey predator (Charnov *et al.* 1976): a monkey may have greater foraging success if it avoids feeding with others whose activities decrease availability of prey. Whatever the reason, it seems clear that invertebrates are a high-priority and valuable resource, one that is worth taking increased risk to obtain.

While the present analysis has focused on protective cover and proximity to conspecifics while feeding, blue monkeys have at least one other major behavioral strategy to protect themselves against predators, namely association in mixed-species groups. Where they coexist with other arboreal monkeys, such associations often occur (Cords 1990b, Struhsaker 1981), though they may vary across sites in their frequency and significance (Chapman and Chapman 2000, Cords 1990b). At Kakamega, blue monkeys associate particularly often with congeneric redtailed monkeys (*C. ascanius*), spending approximately half of their time together; these associations occur more often and last longer than would be expected by chance, and I (Cords 1987, 1990b) have argued that blue and redtailed monkeys are attracted to one another at this site. For blue monkeys, the major benefit of associations with redtails appears to be lower predation risk, and blue monkeys reduced their vigilance when associated with redtails (Cords 1990a).

While association with redtails was not monitored concurrently with the behavior of blue monkey adult females analyzed in this chapter, my previous study on associations between the two species at Kakamega (Cords 1987) provides a few relevant observations. First, blue monkeys at Kakamega are more likely to be feeding when in association with redtails than when in single-species groups: this suggests that the antipredator benefits of associations may be particularly (though not exclusively) applicable to a feeding context. Second, associations were most likely to occur at times of day when the proportion of feeding time devoted to invertebrates was highest. It is thus possible that when they foraged for invertebrates, blue monkeys relied most on the antipredator benefits of associating with redtails, rather than on the cover of foliage or close proximity to members of their own species, as a protective strategy. The two species do differ in the capture methods and substrates used for invertebrate capture (Cords 1987). While these differences are small, they may reduce feeding competition relative to what would occur when only conspecifics are nearby, and may also reduce the likelihood that prey become less available because they are disturbed by the foraging activities of nearby monkeys.

In sum, blue monkeys at Kakamega appeared to reduce the risk of predation while feeding in various ways. Most feeding occurred outside of trees with the lowest foliage density, where monkeys would be most conspicuous to predators, and individuals usually shared their feeding trees with at least one conspecific, with whom vigilance could be shared. The monkeys did feed alone for a large proportion of the time, however, and their feeding trees were seldom filled to capacity. The tendency of these monkeys to space themselves out while feeding may be explained by their avoidance of competition for food, especially invertebrate prey. This evidently important component of their diet was eaten in risky circumstances, namely when protective cover and the potential for shared vigilance were minimal. There was limited evidence that blue monkeys at Kakamega traded off the benefits of foraging in dense foliage with the benefits of foraging near group-mates. Finally, the monkeys often associated with redtail monkeys, and the scheduling of these associations, as well as the way they influenced vigilance behavior, suggest that they also may reduce predation risk during feeding and foraging.

The degree to which the results of this study will hold for blue monkeys at other sites remains to be seen. The nature of one antipredator behavioral response, namely mixed-species association, is already known to vary in occurrence and significance across sites (Chapman and Chapman 2000, Cords 1990b). In particular, it appears that blue monkeys at Kakamega rely more on mixed-species association for protection than do blue monkeys at Kibale (Chapman and Chapman 2000, Cords 1987, 1990b, Struhsaker 1981). Assuming similar predation risk, one might therefore expect that the protection of foliage density or nearby neighbors would be even more important at Kibale. At present, however, comparable data are not available for a throrough comparison of these two sites, or others.

Acknowledgments

My field research has been carried out with permission from the Office of the President and the Ministry of Education, Science and Technology, under sponsorship from the Zoology Department, University of Nairobi, and with the cooperation of local officers of the Forest Department and Kenya Wildlife Service. Fieldwork has been supported by the National Science Foundation (SBR 95–23623, BCS 98–08273), the L.S.B. Leakey Foundation, the Wenner-Gren Foundation and Columbia University. Helpful comments were provided by Lynne Miller and two anonymous reviewers.

REFERENCES

Beeson, M. (1989). Seasonal dietary stress in a forest monkey (*Cercopithecus mitis*). *Oecologia* **78**: 565–70.

Brown, L.H. (1971). The relations of the crowned eagle *Stephanoaetus coronatus* and some of its prey animals. *Ibis* **113**: 240–3.

Brown, L.H., Urban, E.K., and Newman, K. (1982). *The Birds of Africa*, Vol. 1. New York: Academic Press.

Butynski, T.M. (1990). Comparative ecology of blue monkeys (*Cercopithecus mitis*) in high- and low-density subpopulations. *Ecological Monographs* **60**: 1–26.

Chapman, C.A., and Chapman, L.J. (2000). Interdemic variation in mixed-species association patterns: common diurnal primates of Kibale National Park, Uganda. *Behavioral Ecology and Sociobiology* **47**: 129–39.

Charnov, E.L., Orians, G.H., and Hyatt, K. (1976). Ecological implications of resource depression. *American Naturalist* **110**: 247–59.

Cordeiro, N.J. (1992). Behaviour of blue monkeys in the presence of crowned eagles. *Folia Primatologica* **59**: 203–7.

Cords, M. (1986). Interspecific and intraspecific variation in diet of two forest guenons, *Cercopithecus ascanius* and *C. mitis*. *Journal of Animal Ecology* **55**: 811–27.

Cords, M. (1987). *Mixed-species Association of* Cercopithecus *Monkeys in the Kakamega Forest, Kenya*. Berkeley: University of California Press.

Cords, M. (1990a). Vigilance and mixed species association in some East African forest guenons. *Behavioural Ecology and Sociobiology* **26**: 297–300.

Cords, M. (1990b). Mixed-species association of East African guenons: general patterns or specific examples? *American Journal of Primatology* **21**: 101–14.

Cords, M. (1995). Predator vigilance costs of allogrooming in wild blue monkeys. *Behaviour* **132**: 559–69.

Cords, M. (2000a). The grooming partners of immature blue monkeys (*Cercopithecus mitis*). *International Journal of Primatology* **21**: 239–54.

Cords, M. (2000b). Agonistic and affiliative relationships of adult females in a blue monkey group. In: P. Whitehead and C. Jolly, eds., *Old World Monkeys*, Cambridge: Cambridge University Press, pp. 453–79.

Isbell, L.A. (1991). Contest and scramble competition: patterns of female aggression and ranging behavior among primates. *Behavioral Ecology* **2**: 143–55.

Isbell, L.A., Pruetz, J.D., and Young, T.P. (1998). Movements of vervets (*Cercopithecus aethiops*) and patas monkeys (*Erythrocebus patas*) as estimators of food resource size, density, and distribution. *Behavioural Ecology and Sociobiology* **42**: 123–33.

Kaplin, B.A., Munyaligoga, V., and Moermond, T.C. (1998). The influence of temporal changes in fruit availability on diet composition and seed handling in blue monkeys (*Cercopithecus mitis doggetti*). *Biotropica* **30**: 56–71.

Lawes, M.J. (1991). Diet of samango monkey (*Cercopithecus mitis erythrarchus*) in the Cape Vidal dune forest, South Africa. *Journal of Zoology, London* **224**: 149–73.

Lawes, M.J., Henzi, S.P., and Perrin, M.R. (1990). Diet and feeding behaviour of samango monkeys (*Cercopithecus mitis labiatus*) in Ngoye Forest, South Africa. *Folia Primatologica* **54**: 57–69.

Lima, S.L., and Dill, L.M. (1990). Behavioral decisions made under the risk of predation: a review and prospectus. *Canadian Journal of Zoology* **68**: 619–40.

Phillips, K.A. (1995a). Resource patch size and flexible foraging in white-faced capuchins (*Cebus capucinus*). *International Journal of Primatology* **16**: 509–19.

Phillips, K.A. (1995b). Foraging-related agonism in capuchin monkeys (*Cebus capucinus*). *Folia Primatologica* **65**: 159–62.

Rudran, R. (1978). Socioecology of the blue monkeys (*Cercopithecus mitis stuhlmanni*) of the Kibale Forest, Uganda. *Smithsonian Contributions to Zoology* **249**: 1–88.

Struhsaker, T.T. (1981). Polyspecific associations among tropical rain-forest primates. *Zeitschrift für Tierpsychologie* **57**: 268–304.

Struhsaker, T.T., and Leakey, M. (1990). Prey selectivity by crowned haw-eagles on monkeys in the Kibali Forest, Uganda. *Behavioral Ecology and Sociobiology* **36**: 435–43.

Treves, A. (1999). Vigilance and spatial adhesion among blue monkeys. *Folia Primatologica* **70**: 291–4.

van Schaik, C.P. (1989). The ecology of social relationships amongst female primates. In: V. Standen and R.A. Foley, eds., *Comparative Socioecology: The Behavioural Ecology of Humans and Other Mammals*. Oxford: Blackwell Scientific Publications, pp. 5–218.

Zimmerman, D.A., Turner, D.A., and Pearson, D.J. (1996). *Birds of Kenya and Northern Tanzania*. Princeton: Princeton University Press.

14 • Predicting predation risk for foraging, arboreal monkeys

ADRIAN TREVES

Introduction

Predation is among the most important selective pressures on animals. In addition to its direct effects leading to mortality, predation can act indirectly by shaping behavior and ecology (Boinski *et al.* 2000, Isbell 1994, Lima 1998). Often the clearest evidence for indirect action is seen in foraging animals. Many foragers take predators into account when selecting feeding and resting sites (Caraco *et al.* 1980, Cresswell 1994, Ferguson *et al.* 1988, FitzGibbon 1990, 1993, Frid 1997, Holmes 1984, Hughes and Ward 1993, Lima 1987, 1992a, 1993, Werner *et al.* 1983). Similar evidence exists for primates. For example, use of refuges and habitat choice appear to be influenced by predation risk (e.g., Cords 1990, Cowlishaw 1994, 1997, Rose and Fedigan 1995, Treves 1997), but few studies examine primate foraging decisions under the threat of predation.

Understanding how primates make foraging decisions under the threat of predation is important in applied fields as well as theoretical investigations. Consider primate crop raiding. If the severity of crop raiding varies with the risk of predation by humans, farmers could manage their holdings to elevate the real or perceived risk of predation. Optimal group size theory also relies on inferences about foraging under predation risk. Optimal group size is typically modeled as a balance between predation risk and feeding competition within groups (Terborgh and Janson 1986). If joining a large group leads to greater interindividual separation during foraging, risk of predation may increase for foragers in larger groups. In sum, a better understanding of predator sensitive foraging will inform many subfields of primatology.

The approach I take in this chapter is to explore individual vigilance (defined as visual scanning of the surroundings beyond the immediate vicinity) in foraging, arboreal primates. I focus on vigilance behavior because it has long been used to understand the relationship between foraging decisions and predation risk (reviewed in Bednekoff and Lima 1998a, Elgar 1989, Lima 1990, 1992b, Lima and Bednekoff 1999). Each allocation of effort to vigilance detracts from visual search for new food items (Treves 2000), hence feeding rate declines with increased vigilance (Glück 1987, Lendrem 1983, 1984). In turn, vigilance usually increases with the risk of predation (reviewed in Treves 2000), leading animals to feed less efficiently in dangerous areas. I use these well-documented facts to ask if individual vigilance of foraging primates reflects variation in the risk of predation. Variation in the risk of predation is identified indirectly by comparing animals that vary in:

1. access to refuge,
2. protective and obstructive cover, and
3. ease of escape.

Arboreal primates are good subjects for such a study because they travel through a complex, three-dimensional network containing multiple refuges, sources of cover and escape routes. Typically, some refuges are inaccessible to an attacking predator. For example, raptors large enough to kill a monkey are often too large to attack a monkey near the trunk of a tree. (This idea was proposed early on by Freeland 1977 and is consistent with observations by Brown 1966, 1971, Cordeiro 1992, Daneels 1979, Eason 1989, Fowler and Cope 1964, Gautier-Hion and Tutin 1988, Leland and Struhsaker 1993, Peres 1990, Rettig 1978; Overdorff et al., Chapter 8 this volume.) Similarly, most large terrestrial predators such as leopards (*Panthera pardus*) and chimpanzees (*Pan troglodytes*) have difficulty in reaching thin, terminal branch tips (Boesch 1994, Boesch and Boesch 1989, Busse 1980, Cowlishaw 1994, Emmons 1987, Stanford 1996). As a result, accessibility within the canopy may affect risk of predation. This raises the first basic question that requires a test: Does position in the tree affect vigilance levels of foraging monkeys?

Prediction 1: When individuals are positioned in microsites accessible to their major predators, vigilance should increase. I test this prediction in two ways. First, I test if wild red colobus (*Procolobus badius*) have higher vigilance when foraging near the ground than those further up in the canopy, as a response to the risk of chimpanzee attack. Similarly, I test the reverse prediction for redtail monkeys (*Cercopithecus ascanius*), as a response to the risk of attack by raptors.

Prediction 2: Using the same logic as above, I test if red colobus foraging on thin, terminal branch tips are less vigilant than those foraging closer to the trunks of trees, because attacking chimpanzees are more likely to use the trunk as an entry point into the canopy. By contrast, redtail monkeys should have lower vigilance when positioned near the trunk, because raptors seem to pluck prey from terminal branch tips.

In addition to providing refuges, the arboreal environment poses a visual challenge for predators and prey. Predator and prey can pass each other unnoticed if vegetation is dense. Cords (1990) found that wild redtail monkeys and blue monkeys (*Cercopithecus mitis*) decreased vigilance when positioned in dense foliage. Human observers searching for monkeys have more difficulty when animals are foraging in dense foliage (protective cover). However, the vigilance of the monkeys themselves may be impeded if foliage is too dense (obstructive cover). This suggests a bivariate relationship (Lazarus and Symonds 1992), so I pose three basic questions about vigilance and cover without making specific predictions.

1. Does surrounding foliage within reach of a foraging monkey affect its vigilance?
2. Does the foliage density of a tree in which a monkey is foraging affect its vigilance?
3. Do surrounding foliage and tree foliage interact to affect vigilance?

Finally, arboreal monkeys may benefit from early warning to escape predators, as do many other animals (Bednekoff and Lima 1998b, FitzGibbon 1989, Kenward 1978, Lima 1994). If escape is impeded in some way, the animal presumably forages at higher risk. I turn to Belizean black howler monkeys (*Alouatta pigra*) to explore impeded escape. Howler monkeys often use their prehensile tails to feed upside down (personal observation). When finished feeding upside down, a black howler virtually always pulls itself up, rights itself, and then proceeds. This is slower than proceeding from an upright position. Even when alarmed, they were never observed to simply drop downward from an inverted feeding position. As a result, rapid escape appears to be impeded when black howlers forage upside down. They should then benefit from higher vigilance to ensure the earliest warning of danger. Accordingly, I test whether feeding posture affects individual vigilance.

Prediction 3: Black howlers should be more vigilant when feeding upside down than when they feed upright.

Methods

Study Site and Subjects

Monkeys from two sites were studied. The first, Kanyawara in Kibale National Park, western Uganda, is a moist, mid-altitude rainforest with high tree species diversity (Chapman *et al.* 1997, Struhsaker 1997). Across the study site, canopy height varied from 20 m to 50 m and the forest contained many gaps due to natural treefalls and trails of various sizes (Chapman *et al.* 1997). Visibility in the understory rarely exceeded 20 m (unpublished data). The second site, Lamanai Reserve in north-central Belize, is a semi-deciduous subtropical forest regenerating from recent human habitations (Gavazzi 1995, Treves 2001). Across the study site, canopy height ranged from 10 m to 30 m and the forest was characterized by gaps with little continuous canopy above 20 m (unpublished data).

Red colobus are folivore–frugivores that live in stable social groups averaging 50 members (for details of ecology and behavior see Struhsaker 1975, 1980, Struhsaker and Leland 1979, Treves 1998, 1999). Four groups ranging in size from 22 to 76 were studied from 1994 to 1995. Although individual identification was obtained for only a handful of these individuals, there is no evidence of disproportionate sampling of certain individuals (Treves 1998). Chimpanzees prey heavily on red colobus at Kibale, as they do elsewhere in Africa when the two species are sympatric (Boesch 1994, Boesch and Boesch 1989, Bshary and Noë 1997a, Mitani and Watts 1999, Stanford *et al.* 1994, Treves 1999, Uehara *et al.* 1992, Wrangham and Riss 1990). Over 90% of the mammalian prey of chimpanzees ($n = 128$) were red colobus at a site less than 10 km from Kanyawara (Mitani and Watts 1999). Chimpanzee fecal samples from nearby Kanyawara site are consistent with red colobus being the preferred prey (Wrangham *et al.* 1991). Red colobus at three sites respond strongly to real and simulated chimpanzee encounters by increasing time spent scanning and moving higher into the canopy (Bshary and Noë 1997b, Stanford 1995, Treves 1997, 1999).

Redtail monkeys are frugivore–insectivores that live in stable social groups averaging 25 members (for details of ecology and behavior see Cords 1986, 1988, Struhsaker 1980, Struhsaker and Leland 1979, Treves 1998, 1999). Five groups ranging in size from 12 to 29 were studied from 1994 to 1995. This study population was completely sympatric with the red colobus described above. Redtail monkeys responded to several types of predators, typically by increasing interindividual distances and altering patterns of vigilance (Treves 1999). Crowned hawk eagles (*Stephanoaetus coronatus*) are the main predator of redtail monkeys at this site. Nest remains suggest that crowned hawk eagles prey on redtail monkeys slightly more than expected by chance (Struhsaker and Leakey 1990). The only observed predation during my study was caused by a large raptor.

Black howler monkeys are folivore–frugivores that live in stable social groups averaging 5.3 members ($n = 17$: Treves 2001; for details of ecology and behavior see Horwich 1983, Horwich and Gebhard 1983, Silver *et al.* 1998, Treves 2001, Treves *et al.* 2001). Six groups of black howlers were studied from 1997 to 2000 but only 12 individuals provided sample of vigilance during foraging in both upright and inverted postures. All study subjects were individually identifiable by facial features, coloration and markings. Although we have seen no successful predation on howlers at Lamanai, potential mammalian predators include tayras (*Eira barbara*), jaguarundis (*Felis yaguarondi*), pumas (*F. concolor*), and jaguars (*Panthera onca*) (Emmons 1987, Galef *et al.* 1976, Peetz *et al.* 1992, Schaller 1983). Potential aerial predators include ornate hawk eagles (*Spizaetus ornatus*), and crested eagles (*Morphnus guianensis*) (Julliot 1994, Rettig 1978; M. England personal communication 1999).

Data collection

Vigilance was sampled using continuous focal animal samples (1 min in length for red colobus and redtail monkeys, 2 min in length for black howlers). This technique captures the briefest of glances (Metcalfe 1984). Time spent scanning is analyzed as a percentage (SCAN). In most vigilance studies, an animal with its head up and eyes open is scored as vigilant (Treves 2000). For arboreal monkeys, whose attackers may come from below, I defined vigilance as any scanning directed beyond arm's reach of the focal animal. This included glances to associates if positioned beyond arm's reach, and excluded glances toward the focal animal's own body, a grooming partner, nearby substrate or food item. Even if a monkey surveys a travel route or distant food, scanning beyond arm's reach may reveal a threat. The need to differentiate scans beyond arm's reach from those directed within arm's reach required that I see the eyes of the focal animal clearly. If the focal animal's eyes were obscured for more than 2 sec, or the animal traveled more than 10 m or between trees, I aborted the sample. Therefore, my data consist of records collected in good visibility from slow-moving or stationary animals.

Following a successful vigilance sample, the context of the focal animal was noted. Here individual class was categorized as adult female, adult male or immature, based on physical features (CLASS). Along with CLASS, the group size and number of neighbors within 2 m (NEIGHBOR) were also noted. Group size did not predict vigilance for any of the three species (Treves 1998, Treves et al. 2001). NEIGHBOR was categorized dichotomously as none (no neighbors within 2 m) or some (one or more). For both red colobus and redtail monkeys, only one age–sex class proved to reduce vigilance when associates were within 2 m (Treves 1998) so I searched for interactions between CLASS and NEIGHBORS. For black howlers, both sexes showed the pattern (Treves et al. 2001).

For red colobus and redtail monkeys, the microsite of the focal animal was described as follows. Height from the ground was estimated in meters. Exercises in interobserver reliability and calibration of these estimates were run twice. For use in ANOVA (analysis of variance) tests, height in m is broken into 10–m blocks (HEIGHT). Distance from the trunk of the tree (TRUNK POSITION) was recorded with three categories: 'Distal' referred to focal animals positioned within 2 m of the terminal branch tip. 'Trunk' referred to a focal animal positioned within 2 m of the main vertical stem of the tree. 'Medial' referred to all other positions. Two distinct measures of cover were used. Surrounding foliage density (SFD) was estimated using a technique similar to that of Cords (1990). SFD ranged from 0 to 2. To code it, a cube was imagined surrounding the focal animal at its arm's length. The observer counted the number of cube faces that were more than 50% obscured by vegetation. If less than two obscured cube faces surrounded the focal animal, then SFD was assigned a zero (i.e., the animal was highly visible). If two or three cube faces were obscured, then SFD equaled 1. If four or more faces were obscured, then SFD equaled 2. Tree foliage density (TFD) was estimated by multiplying the average surface area of mature leaves by the number of mature leaves on branch tips. Five trees of each species were selected at random to estimate TFD. The observer first counted all mature leaves within 1 m of the tips of ten branches in the middle canopy (approximately horizontal branches). These ten values (LEAF #) were then averaged and multiplied by LEAF SIZE, estimated as follows. From each tree used above, ten fallen, intact, mature leaves were collected. The major and minor axis of each leaf was measured in millimetres, treating each leaf as if it were elliptical in shape. These axes values were then multiplied to derive the average square area of leaf (LEAF SIZE). Both LEAF # and LEAF SIZE are reported as an average of five individual trees for each species, and their product is the index TFD. The tree species analyzed here are those – among the top ten eaten by redtail monkeys and red colobus – for which adequate samples of vigilance were available (more than ten in each species of tree). Trees were identified and measured by P. Baguma and F. Mugurusi.

Activity of the focal animal was noted at the end of each sample (foraging, resting, other). Only foraging animals are considered here and the food type (FOOD) they ate was noted (flower, fruit, insect, leaf, other). For red colobus and redtail monkeys, the single activity observed at the end of 1-min sample was noted. This introduces the possibility that a foraging animal had begun by resting and ended by foraging. I cannot estimate the magnitude of this error. The technique was refined in the study of black howlers by recording all activities engaged in throughout the 2-min interval. An animal that engaged in any of the following nonforaging activities (resting, moving, socializing) was excluded from the present analyses. Hence, only black howlers that engaged exclusively in foraging were analyzed here. Also, black howlers foraging upside down were distinguished from those foraging right side up (POSTURE).

Results

To test if the vigilance of foraging monkeys differed by microsite, I analyzed red colobus and redtail monkeys separately. For both species, I used an ANOVA to control simultaneously for CLASS, FOOD, NEIGHBOR, and their pairwise interactions, while testing microsite factors HEIGHT, TRUNK POSITION, and SFD.

For foraging red colobus, CLASS ($F_{2,466}=4.21$, $p=0.015$), NEIGHBOR ($F_{1,466}=6.73$, $p=0.0098$), and the interaction of the two ($F_{2,466}=5.58$, $p=0.0040$) were all significant. The significance of CLASS arose only because adult females were less vigilant than adult males (post hoc Fisher's PLSD (protected least significant difference) $p=0.0001$). Focal animals with associates within 2 m were less vigilant than those without and adult males displayed the pattern most strikingly, as in Treves (1998). FOOD ($F_{4,466}=1.85$, $p=0.12$) and its interactions ($p>0.30$ for both) were not significant. Controlling for the

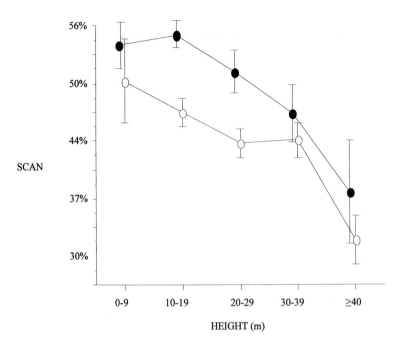

Fig. 14.1. Percent of time spent scanning (SCAN) in relation to height in the canopy for foraging, arboreal monkeys. Open circles are red colobus, while solid circles (•) are red-tail monkeys. Error bars depict 1 SEM (standard error of the mean).

significant factors described above, red colobus near the ground were more vigilant than those higher up (HEIGHT: $F_{4,466} = 5.51$, $p = 0.0002$). Neither TRUNK POSITION ($F_{2,466} = 1.53$, $p = 0.22$), nor SFD ($F_{2,466} = 0.86$, $p = 0.43$) were informative.

For foraging redtail monkeys, I found no significant effect of CLASS ($F_{2,336} = 0.77$, $p = 0.46$), NEIGHBOR ($F_{1,336} = 0.11$, $p = 0.74$), their interaction ($F_{2,336} = 2.39$, $p = 0.094$), nor FOOD ($F_{4,336} = 0.34$, $p = 0.85$). As for microsite, redtails were also more vigilant lower in the canopy (HEIGHT: $F_{4,336} = 3.39$, $p = 0.0097$), and closer to the trunk ($F_{2,336} = 3.63$, $p = 0.028$). SFD was not significant ($F_{2,336} = 2.06$, $p = 0.13$).

In sum, both red colobus and redtails were more vigilant near the ground (Fig. 14.1). Both species were more vigilant when positioned near the trunk (Fig. 14.2), although this was not significant for red colobus. Neither species responded to surrounding foliage density (Fig. 14.3).

Tree foliage density (TFD)

LEAF #, LEAF SIZE and TFD for 12 tree species are presented in Table 14.1 along with average SCAN (pooled across species). The analysis was run in three steps. First, I computed an ANOVA incorporating

Fig. 14.2. Percent of time spent scanning (SCAN) in relation to distance from the trunk for foraging, arboreal monkeys. Open circles are red colobus, while solid circles (•) are redtail monkeys. Error bars depict 1 SEM.

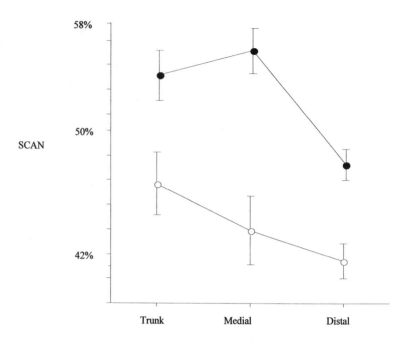

Fig. 14.3. Percent of time spent scanning (SCAN) in relation to the density of foliage surrounding (SFD) foraging, arboreal monkeys. Open bars are red colobus,while solid bars are redtail monkeys. Error bars depict 1 SEM.

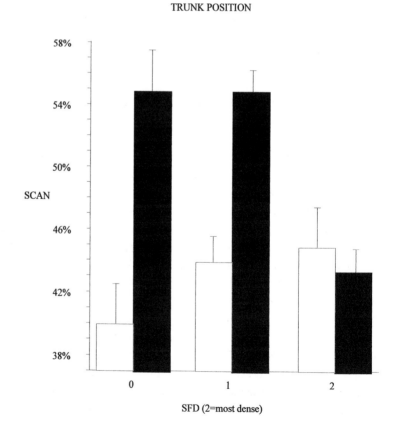

Table 14.1. *Tree foliage densities of the most common food trees of red colobus and redtail monkeys at Kibale*

| Tree species | Tree foliage density | | | Vigilance | |
	LEAF #	LEAF SIZE	TFD index	n	mean $\pm$ SE (%)
Celtis africana	41.5 $\pm$ 3.3	34.4 $\pm$ 5.1	1430.6	15	26.5 $\pm$ 5.2
Symphonia globulifera	30.5 $\pm$ 1.9	47.6 $\pm$ 1.2	1452.4	13	52.8 $\pm$ 5.0
Ficus brachylepis	31.5 $\pm$ 6.7	47.6 $\pm$ 1.2	2328.0	31	41.8 $\pm$ 3.3
Parinari excelsa	35.7 $\pm$ 1.6	76.5 $\pm$ 6.2	2732.5	26	33.3 $\pm$ 4.5
Chrysophyllum gorungosanum	24.8 $\pm$ 1.7	104.2 $\pm$ 24.3	2960.4	27	24.7 $\pm$ 3.3
Ficus exasperata	23.4 $\pm$ 2.1	127.0 $\pm$ 22.7	2973.9	16	57.7 $\pm$ 6.7
Markhamia platycalyx	19.8 $\pm$ 2.4	174.6 $\pm$ 26.6	3452.8	28	50.8 $\pm$ 3.7
Pygium africanum	38.8 $\pm$ 5.5	89.1 $\pm$ 7.0	3461.4	24	53.7 $\pm$ 4.3
Celtis durandi	30.7 $\pm$ 4.8	122.3 $\pm$ 10.8	3653.4	36	47.7 $\pm$ 3.0
Albizia gummifera	26.6 $\pm$ 4.1	171.2 $\pm$ 10.4	4556.3	21	55.8 $\pm$ 5.0
Funtumia latifolia	19.2 $\pm$ 1.7	248.6 $\pm$ 11.3	4767.4	13	40.3 $\pm$ 6.0
Strombosia scheffleri	20.8 $\pm$ 0.8	282.6 $\pm$ 19.9	5884.1	41	55.8 $\pm$ 3.3

Notes:

LEAF #: the number of mature leaves within 1 m of branch tip (average of 50 branches from five trees); LEAF SIZE: the area in square millimeters of fallen, mature leaves (average of 50 leaves from five trees); TFD index: LEAF multiplied by LEAF SIZE.

monkey species, CLASS, NEIGHBOR, the interaction of the latter two, FOOD, three microsite variables (HEIGHT, TRUNK POSITION and SFD), and finally tree species. This permitted me to ask if tree species had an effect on vigilance. Tree species was recorded in only 374 samples, hence this analysis had lower power than previous ones. Nevertheless, tree species was strongly significant ($F_{17,320} = 3.85$, $p = 0.0001$), along with CLASS ($F_{2,320} = 7.95$, $p = 0.0004$), NEIGHBOR $\times$ CLASS ($F_{2,320} = 4.42$, $p = 0.013$), FOOD ($F_{3,320} = 3.66$, $p = 0.013$) and SFD ($F_{2,320} = 3.60$, $p = 0.028$). This result suggests that tree species affects the vigilance of foraging monkeys but it leaves open the question of whether tree foliage density (TFD) can explain the differences, or some other feature that differs between tree species. Interestingly, monkey species was insignificant ($F_{1,320} = 0.004$, $p = 0.95$), thus I pool the species for the remainder of this analysis.

The next step was to replace tree species with TFD (Table 14.1) and the interaction of SFD and TFD, and then exclude the insignificant variables from the preceding analysis. For use in the ANOVA, TFD was split into categories of high and low. There were six tree species

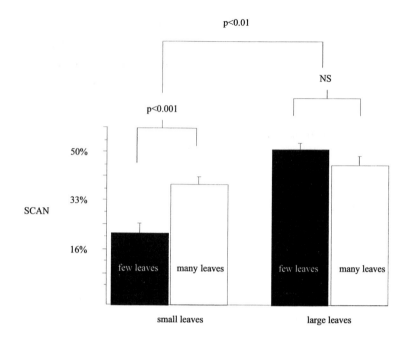

Fig. 14.4. Percent of time spent scanning (SCAN) in relation to the size of leaves and number of leaves in selected food tree species (see Table 14.1). Red colobus and red-tail monkeys are pooled. Error bars depict 1 SEM.

with low TFD indices (1430–2973) and six with high (3452–5884). I then repeated the ANOVA. All of the variables identified above remained significant, except SFD ($F_{2,264}=0.39$, $p=0.68$), which seemed to be replaced by a strong effect of TFD ($F_{1,264}=20.10$, $p=0.0001$). The interaction term was not significant ($F_{2,264}=2.67$, $p=0.071$).

Because TFD was computed from LEAF SIZE and LEAF # (Table 14.1), my final step was to determine whether the number of leaves or the size of leaves or some interaction was most responsible for the strong effect of TFD. As before, LEAF # and SIZE were split into two categories of equal size (LEAF # 19.2–28.4 vs. 30.5–41.5 and LEAF SIZE 34.4–104.2 vs. 122.3–282.6). The final ANOVA revealed that both LEAF SIZE ($F_{1,267}=5.59$, $p=0.019$) and the interaction of LEAF SIZE and LEAF # ($F_{1,266}=4.26$, $p=0.04$) were significant (Fig. 14.4), although neither approached the strength of the composite index (TFD). Focal animals of both species were least vigilant in trees with few, small leaves and vigilance was significantly higher when trees had large leaves.

Foraging posture

Two analyses were used to test if black howler monkeys were more vigilant when foraging upside down. First, I used an ANOVA incorporating CLASS, FOOD, and POSTURE plus the pairwise interaction of each. Only POSTURE was significant ($F_{1,67} = 9.0$, $p = 0.004$), with upside down animals spending an average of 32% (SE 5.2) of their time scanning and upright animals spending 47.4% (SE 4.1). I then took advantage of individual identification in this study to exert more control over variation. I used a Wilcoxon signed-ranks test to compare each individual's vigilance when foraging upside down to that when foraging upright. Twelve black howlers contributed to this analysis and 11 of the 12 had higher vigilance when foraging upright ($16.8 \pm 5.6\%$ more time spent scanning when upright: $Z = 2.51$, $p = 0.012$).

Discussion

The premise of this chapter is that risk of predation affects vigilance which, in turn, affects foraging rate. Specifically, I assumed that foraging monkeys exposed near the ground, near branch tips, in sparse foliage, or those foraging upside down, faced a higher risk of predation. I then went on to make predictions – about specific predator–prey pairs – that vigilance would be higher in more vulnerable monkeys to counter risk. Most of the predictions were rejected.

Height in the forest canopy predicted the vigilance levels of both red colobus and redtail monkeys in similar fashion (Fig. 14.1). Red colobus had higher vigilance when foraging near the ground, as predicted. This is consistent with avoidance of chimpanzees, the major predator of red colobus across Africa. Contrary to prediction 1, redtail monkeys also had higher vigilance when foraging near the ground. The opposite pattern was predicted because raptors are believed to pose the major predatory threat to Kibale's redtail monkeys (Struhsaker and Leakey 1990). Either this assumption is wrong, or both species are behaving consistently according to some other criterion. I present another explanation in the section below, entitled 'Obstructive cover.'

Red colobus and redtail monkeys responded similarly to trunk position. Both species tended to be less vigilant when positioned near a branch tip (Fig. 14.2). Red colobus vigilance was not significantly lower at distal positions, so prediction 2 is rejected unequivocally. Again the similarity of the two species argues for a consistent and general antipredator explanation, rather than one based on specific interactions of particular predator–prey pairs.

Neither red colobus nor redtail monkeys responded to the

density of foliage surrounding foragers, when confounding variables – such as food type – were considered (Fig. 14.3). At a larger scale, both monkey species were less vigilant in trees with small leaves and particularly those with few, small leaves (Fig. 14.4). I found no interaction between foliage density of the tree and that immediately surrounding a foraging monkey. My index of tree foliage density near branch tips – though crude – emerged as the single most powerful effect on arboreal monkey vigilance. This suggests further work is needed to understand foliage density and cover (see below).

For black howler monkeys, I found that upright animals were more vigilant than those up-side down. This contradicted expectations based on ease of escape.

Refuge, cover and ease of escape

The predictions laid out in the Introduction fared very poorly. The logic underlying these predictions, although simple and intuitively sensible, could not explain the vigilance patterns of the arboreal primates I studied. I offer three possible explanations.

First, the assumption that vigilance will be higher when predation risk is higher may be incorrect (see for example Lima 1987). Indeed, I previously reported that redtail monkeys did not increase time spent scanning when exposed to sound playbacks of potential predators. However, they did alter their vigilance patterns in ways that suggested they focused attention on the likely avenues of approach by those predators (Treves 1999). Moreover, dozens of studies of primates and nonprimates alike report that vigilance increases with predation risk (Treves 2000). Because formulating special predictions for each species of primate is not intuitively satisfying nor parsimonious, I see no reason at present to reject the assumption that vigilance increases with the risk of predation.

Alternatively, my assumptions about predation risk may be faulty. Perhaps red colobus and redtail monkeys are safer when they are high in the canopy, positioned at terminal branch tips or located in sparse vegetation. These microsites may facilitate escape (e.g., Gebo *et al.* 1994). Likewise, black howlers may be safer when foraging upside down (or may do so only when it is safe). Although my assumptions about risk were based on field observations of both predators and prey, we will need experimental tests to confirm them. In particular, we need to determine if risky positions are adopted primarily when animals feel safe (low risk).

A third alternative is possible. I may have overstated the importance of microsite and posture. Perhaps predictions based on particular predator–prey pairs (e.g., redtails and raptors) are too specific. A more powerful force shaping vigilance in the arboreal monkeys I studied may be variation in the probability of detecting predators.

Obstructive cover

From studies of nonprimates, we know that obstructive cover elevates vigilance, because more time and effort are needed to detect predators when visibility is reduced (Lazarus and Symonds 1992, Metcalfe 1984, Roberts 1996). Presumably, foraging primates try to detect threats before they close. Therefore, arboreal monkeys may vary their vigilance levels according to the ease of predator detection *not* according to their own exposure or vulnerability.

Many of the results of this study can be interpreted parsimoniously as a response to obstructive cover. Monkeys low in the canopy contend with low light, dense understory vegetation and a lower vantage point. Under these conditions, both species showed higher vigilance (Fig. 14.1), regardless of any particular predator. Similarly, animals near the trunk face a concentration of large branches and the surrounding crown of their tree. Both species tended to spend more time scanning when positioned at the trunk (Fig. 14.2). The absence of an effect of surrounding foliage density can also be interpreted in this way (Fig. 14.3). Denser foliage would seem to obstruct vigilance and hence increase effort. However, observers could only sample vigilance when they saw the face of the focal animal, hence the foliage immediately surrounding a focal animal may not obstruct its own vigilance. Tree foliage fits this interpretation more clearly. Animals foraging in trees with large leaves or many, smaller leaves were screened from their surroundings, leading them to devote more time to vigilance (Fig. 14.4). For black howler monkeys foraging in an inverted posture, the role of obstructive cover is not obvious. Are inverted black howlers better able to see threats from below, or are they instead confronted by such an obstruction (the ground) that they reduce attempts to scan? Perhaps upright animals have more to survey and need to devote more time to do so. This conjecture awaits further research, but we should expect black howlers to take other measures to limit risk while foraging upside down, such as abbreviating the time spent in this position (Lima 1987). In sum, explanations based on visibility and obstructive cover are effective in predicting patterns of vigilance, and they are simpler, because they do not rely on species-specific responses to particular predators.

The logic laid out in the Introduction assumes that predators primarily capture prey when the prey are accessible. A competing explanation is that predators primarily capture prey that are unaware. The difficulty of detecting predators may determine individual risk of predation more powerfully than accessibility.

If visibility is more important than accessibility to predators, a set of common assumptions about predation risk will need to be verified for primates. For example, predation risk is often assumed to increase in open-country habitats (e.g., Anderson 1986, Hill and Lee 1998), yet the detection of predators may be easier in these habitats because prey face less obstructive cover. Similarly, nocturnal activity and the ability of nocturnal forms to travel alone have both been attributed to a lower risk of predation at night (Terborgh and Janson 1986, van Schaik and van Hooff 1983). Yet, low light may impede the visual detection of predators by prey, thereby elevating risk. This will be particularly true for predators with keen night vision. Our predictions about predation risk should rest on the relative sensory acuity of predator and prey, not on our human sensibilities. Greater attention to visibility, obstructive cover, and most importantly, the ease with which primates detect their predators may advance our understanding of predator–prey coevolution in primates.

Acknowledgments

I wish to thank P. Baguma, A. Drescher, N. Ingrisano, F. Mugurusi, L. Naughton and R. Wrangham for help and support in the field. The author was supported by the University of Wisconsin-Madison during the writing. Permission from the Ugandan National Research Council and the Belizean Department of Forestry are also gratefully noted. Useful comments from L. Miller and two anonymous reviewers helped strengthen the manuscript.

REFERENCES

Anderson, C.M. (1986). Predation and primate evolution. *Primates* **27**: 15–39.
Bednekoff, P.A., and Lima, S.L. (1998a). Randomness, chaos and confusion in the study of antipredator vigilance. *Tree* **13**: 284–7.
Bednekoff, P.A., and Lima, S.L. (1998b). Re-examining safety in numbers: Interactions between risk dilution and collective detection depend

upon predator targeting behaviour. *Proceedings of the Royal Society of London Series B* **265**: 2021–26.

Boesch, C. (1994). Chimpanzees–red colobus monkeys: A predator–prey system. *Animal Behaviour* **47**: 1135–48.

Boesch, C., and Boesch, H. (1989). Hunting behavior of wild chimpanzees in the Tai National Park. *American Journal of Physical Anthropology* **78**: 547–73.

Boinski, S., Treves, A., and Chapman, C.A. (2000). A critical evaluation of the influence of predators on primates: Effects on group movement. In: S. Boinski and P. Garber, eds., *On the Move: How and Why Animals Travel in Groups*, Chicago: University of Chicago Press, pp. 43–72.

Brown, L.H. (1966). Observations on some Kenya eagles. *Ibis* **108**: 531–72.

Brown, L.H. (1971). The relations of the crowned eagle *Stephanoaetus coronatus* and some of its prey animals. *Ibis* **113**: 240–3.

Bshary, R., and Noë, R. (1997a). Red colobus and diana monkeys provide mutual protection against predators. *Animal Behaviour* **54**: 1461–74.

Bshary, R., and Noë, R. (1997b). Anti-predation behaviour of red colobus monkeys in the presence of chimpanzees. *Behavioral Ecology and Sociobiology* **41**: 321–33.

Busse, C. (1980). Leopard and lion predation upon chacma baboons living in the Moremi Wildlife Reserve. *Botswana Notes & Records* **12**: 15–20.

Caraco, T., Martindale, S., and Pulliam, H.R. (1980). Avian time budgets and distance to cover. *The Auk* **97**: 872–5.

Chapman, C.A., Chapman, L.J., Wrangham, R., Isabirye-Basuta, G., and Ben-David, K. (1997). Spatial and temporal variability in the structure of a tropical forest. *African Journal of Ecology* **35**: 287–302.

Cordeiro, N.J. (1992). Behaviour of blue monkeys (*Cercopithecus mitis*) in the presence of crowned eagles (*Stephanoaetus coronatus*). *Folia Primatologica* **59**: 203–7.

Cords, M. (1986). Interspecific and intraspecific variation in diet of two forest guenons, *Cercopithecus ascanius* and *C. mitis*. *Journal of Animal Ecology* **55**: 811–27.

Cords, M. (1988). Mating systems of forest guenons: A preliminary review. In: A. Gautier-Hion, F. Bourliere, J.-P. Gautier and J. Kingdon, eds., *A Primate Radiation: Evolutionary Biology of the African Guenons*, Cambridge: Cambridge University Press, pp. 323–39.

Cords, M. (1990). Vigilance and mixed-species association of some East African forest monkeys. *Behavioral Ecology and Sociobiology* **26**: 297–300.

Cowlishaw, G. (1994). Vulnerability to predation in baboon populations. *Behaviour* **131**: 293–304.

Cowlishaw, G. (1997). Refuge use and predation risk in a desert baboon population. *Animal Behaviour* **54**: 241–53.

Cresswell, W. (1994). Flocking is an effective anti-predation strategy in redshanks, *Tringa totanus*. *Animal Behaviour* **47**: 433–42.

Daneels, A.B.C. (1979). Prey size and hunting methods of the crowned eagle. *Ostrich* **50**: 120–1.

Eason, P. (1989). Harpy eagle attempts predation on adult howler monkey. *The Condor* **91**: 469-70.

Elgar, M.A. (1989). Predator vigilance and group size in mammals and birds: A critical review of the empirical evidence. *Biological Review* **64**: 13-33.

Emmons, L.J. (1987). Comparative feeding ecology of felids in a Neotropical rainforest. *Behavioral Ecology and Sociobiology* **20**: 271-83.

Ferguson, S.H., Bergerud, A.T., and Ferguson, R. (1988). Predation risk and habitat selection in the persistence of a remnant caribou population. *Behavioral Ecology and Sociobiology* **76**: 236-45.

FitzGibbon, C.D. (1989). A cost to individuals with reduced vigilance in groups of Thomson's gazelles hunted by cheetahs. *Animal Behaviour* **37**: 508-10.

FitzGibbon, C.D. (1990). Anti-predator strategies of immature Thomson's gazelles: Hiding and the prone response. *Animal Behaviour* **40**: 846-55.

FitzGibbon, C.D. (1993). Antipredator strategies of female Thomson's gazelles with hidden fawns. *Journal of Mammalogy* **75**: 758-62.

Fowler, J.M., and Cope, J.B. (1964). Notes on the harpy eagle in British Guiana. *The Auk* **81**: 257-73.

Freeland, W.J. (1997). Dynamics of primate parasites. Ph.D, thesis, University of Pennsylvania.

Frid, A. (1997). Vigilance by female Dall's sheep: Interactions between predation risk factors. *Animal Behaviour* **53**: 799-808.

Galef, B.G., Mittermeier, R.A., and Bailey, R.C. (1976). Predation by the tayra (*Eira barbara*). *Journal of Mammalogy* **57**: 760-1.

Gautier-Hion, A., and Tutin, C.E.G. (1988). Simultaneous attack by adult males of a polyspecific troop of monkeys against a crowned hawk eagle. *Folia Primatologica* **51**: 149-51.

Gavazzi, A.J. (1995). Ecology of the black howler monkey (*Alouatta pigra*) at Lamanai, Belize. Masters thesis, San Francisco State University.

Gebo, D.L., Chapman, C.A., Chapman, L.J., and Lambert, J. (1994). Locomotor responses to predator threat in red colobus monkeys. *Primates* **35**: 219-23.

Glück, E. (1987). An experimental study of feeding, vigilance and predator avoidance in a single bird. *Oecologia* **71**: 268-72.

Hill, R.A., and Lee, P.C. (1998). Predation risk as an influence on group size in cercopithecoid primates: Implications for social structure. *Journal of Zoology (London)* **245**: 447-56.

Holmes, W.G. (1984). Predation risk and foraging behavior of the hoary marmot in Alaska. *Behavioral Ecology and Sociobiology* **15**: 293-301.

Horwich, R.H. (1983). Breeding behavior in the black howler monkey (*Alouatta pigra*) of Belize. *Primates* **24**: 222-30.

Horwich, R.H., and Gebhard, K. (1983). Roaring rhythms in black howler monkeys (*Alouatta pigra*) of Belize. *Primates* **24**: 290-6.

Hughes, J.J., and Ward, D. (1993). Predation risk and distance to cover affect foraging behaviour in Namib Desert gerbils. *Animal Behaviour* **46**: 1243-5.

Isbell, L.A. (1994). Predation on primates: Ecological patterns and evolutionary consequences. *Evolutionary Anthropology* **3**: 61–71.

Julliot, C. (1994). Predation of a young spider monkey (*Ateles paniscus*) by a crested eagle (*Morphnus guianensis*). *Folia Primatologica* **63**: 75–7.

Kenward, R.E. (1978). Hawks and doves: factors affecting success and selection in goshawk attacks on woodpigeons. *Journal of Animal Ecology* **47**: 449–60.

Lazarus, J., and Symonds, M. (1992). Contrasting effects of protective and obstructive cover on avian vigilance. *Animal Behaviour* **43**: 519–21.

Leland, L., and Struhsaker, T.T. (1993). Teamwork tactics. *Natural History* **4**: 43–8.

Lendrem, D.W. (1983). Predation and risk in the blue tit (*Parus caeruleus*). *Behavioral Ecology and Sociobiology* **14**: 9–13.

Lendrem, D.W. (1984). Flocking, feeding and predation risk: Absolute and instantaneous feeding rates. *Animal Behaviour* **32**: 298–9.

Lima, S.L. (1987). Distance to cover, visual obstructions, and vigilance in house sparrows. *Behaviour* **102**: 231–8.

Lima, S.L. (1990). The influence of models on the interpretation of vigilance. In: M. Bekoff and D. Jamieson, eds., *Interpretation and Explanation in the Study of Animal Behavior*, Boulder: Westview Press, pp. 246–67.

Lima, S.L. (1992a). Vigilance and foraging substrate: anti-predatory considerations in a non-standard environment. *Behavioral Ecology and Sociobiology* **30**: 283–9.

Lima, S.L. (1992b). Life in a multi-predator environment: Some considerations for anti-predatory vigilance. *Ann Zoolici Fennici*, **29**: 217–26.

Lima, S.L. (1993). Ecological and evolutionary perspectives on escape from predatory attack: A survey of North American birds. *Wilson Bulletin* **105**: 1–47.

Lima, S.L. (1994). On the personal benefits of anti-predatory vigilance. *Animal Behaviour* **48**: 734–6.

Lima, S.L. (1998). Nonlethal effects in the ecology of predator–prey interactions. *BioScience* **48**: 25–34.

Lima, S.L., and Bednekoff, P.A. (1999). Temporal variation in danger drives anti-predator behavior: The predation risk allocation hypothesis. *American Naturalist* **153**: 649–59.

Metcalfe, N.B. (1984). The effect of habitat on the vigilance of shorebirds: Is visibility important? *Animal Behaviour* **32**: 981–5.

Mitani, J.C., and Watts, D.P. (1999). Demographic influences on the hunting behavior of chimpanzees. *American Journal of Physical Anthropology* **109**: 439–54.

Peetz, A.A., Norconk, M.A., and Kinzey, W.G. (1992). Predation by jaguar on howler monkeys (*Alouatta seniculus*) in Venezuela. *American Journal of Primatology* **28**: 223–8.

Peres, A. (1990). A harpy eagle successfully captures an adult male red howler monkey. *Wilson Bulletin* **102**: 560–1.

Rettig, N.L. (1978). Breeding behavior of the harpy eagle (*Harpia harpyja*). *The Auk* **95**: 629–43.

Roberts, G. (1996). Why individual vigilance declines as group size increases. *Animal Behaviour* **51**: 1077–86.

Rose, L.M., and Fedigan, L.M. (1995). Vigilance in white-faced capuchins, *Cebus capucinus*, in Costa Rica. *Animal Behaviour* **49**: 63–70.

Schaller, G.B. (1983). Mammals and their biomass on a Brazilian ranch. *Arquivos de Zoologia* **31**: 1–36.

Silver, S.C., Ostro, L.E.T., Yeager, C.P., and Horwich, R. (1998). Feeding ecology of the black howler monkey (*Alouatta pigra*) in northern Belize. *American Journal of Primatology* **45**: 263–79.

Stanford, C.B. (1995). The influence of chimpanzee predation on group size and anti-predator behaviour in red colobus monkeys. *Animal Behaviour* **49**: 577–87.

Stanford, C.B. (1996). The hunting ecology of wild chimpanzees: Implications for the evolutionary ecology of Pliocene hominids. *American Anthropologist* **98**: 96–113.

Stanford, C.B., Wallis, J., Matama, H., and Goodall, J. (1994). Patterns of predation by chimpanzees on red colobus monkeys in Gombe National Park, Tanzania 1982–1991. *American Journal of Physical Anthropology* **94**: 213–28.

Struhsaker, T.T. (1975). *The Red Colobus Monkey*. Chicago: University of Chicago Press.

Struhsaker, T.T. (1980). Comparison of the behaviour and ecology of red colobus and redtail monkeys in the Kibale Forest, Uganda. *African Journal of Ecology* **18**: 33–51.

Struhsaker, T.T. (1997). *Kibale: A Case Study of Logging in a Tropical Rainforest*. Gainesville: University of Florida Press.

Struhsaker, T.T., and Leakey, M. (1990). Prey selectivity by crowned hawk-eagles on monkeys in the Kibale Forest, Uganda. *Behavioral Ecology and Sociobiology* **26**: 435–43.

Struhsaker, T.T., and Leland, L. (1979). Socioecology of five sympatric monkey species in the Kibale Forest, Uganda. In: J.S. Rosenblatt, R.A. Hinde, C. Beer and M.C. Busnel, eds., *Advances in the Study of Behavior*, New York: Academic Press, pp. 159–228.

Terborgh, J., and Janson, C.H. (1986). The socioecology of primate groups. *Annual Review of Ecology and Systematics* **17**: 111–35.

Treves, A. (1997). Vigilance and use of micro-habitat in solitary rainforest mammals. *Mammalia* **61**: 511–25.

Treves, A. (1998). The influence of group size and near neighbors on vigilance in two species of arboreal primates. *Behaviour* **135**: 453–82.

Treves, A. (1999). Has predation shaped the social systems of arboreal primates? *International Journal of Primatology* **20**: 35–53.

Treves, A. (2000). Theory and method in studies of vigilance and aggregation. *Animal Behaviour* **60**: 711–22.

Treves, A. (2001). Reproductive consequences of variation in the composition of howler monkey groups. *Behavioral Ecology and Sociobiology* **50**: 61–71

Treves, A., Drescher, A., and Ingrisano, N. (2001). Vigilance and aggregation in black howler monkeys (*Alouatta pigra*). *Behavioral Ecology and Sociobiology* **50**: 90–95.

Uehara, S., Nishida, T., Hamai, M., Hasegawa, T., Hayaki, H., Huffman, M.A., Kawanaka, K., Kobayashi, S., Mitani, J.C., Takahata, Y., Takasaki, H., and Tsukahara, T. (1992). Characteristics of predation by the chimpanzees in the Mahale Mountains National Park, Tanzania. In: N. Itoigawa, Y. Sugiyama, G.P. Sackett and R.K.R. Thompson, eds., *Topics in Primatology: Behavior, Ecology and Conservation*, Tokyo: University of Tokyo Press, pp. 143–58.

van Schaik, C.P., and van Hooff, J.A.R.A.M. (1983). On the ultimate causes of primate social systems. *Behaviour* **85**: 91–117.

Werner, E.E., Gilliam, J.F., Hall, D.J., and Mittelbach, G.G. (1983). An experimental test of the effects of predation risk on habitat use in fish. *Ecology* **64**: 1540–8.

Wrangham, R.W., and Riss, E.V.Z.B. (1990). Rates of predation on mammals by Gombe chimpanzees, 1972–1975. *Primates* **31**: 157–70.

Wrangham, R.W., Conklin, N.L., Chapman, C.A., and Hunt, K.D. (1991). The significance of fibrous foods for Kibale Forest chimpanzees. *Philosophical Transactions of the Royal Society of London, Series B* **334**: 171–8.

15 • Predator sensitive foraging in ateline primates

ANTHONY DI FIORE

Introduction

Many researchers have assumed that the risk of predation is a major factor – perhaps the most important factor – favoring sociality in primates (Alexander 1974, Dunbar 1988, Terborgh 1983, Terborgh and Janson 1986, van Schaik 1983). Testing this assumption has proven difficult because actual cases of predation on primates are rarely observed (Cheney and Wrangham 1987, Isbell 1994). None the less, if predation is a significant cause of mortality in contemporary primate populations, we would expect individual animals to be sensitive to the *risk of predation* and to adjust their behaviors in response to changing perceptions of that risk. We would particularly expect to see adjustments in the realm of foraging behavior – in decisions over where or when or for how long to forage, or over how vigilant to be while foraging – because the trade-off between avoiding predation and acquiring food is presumed to be fundamental for most taxa (Fraser and Huntingford 1986, Lima and Dill 1990). Although this trade-off has been studied extensively in many nonprimates (reviewed in Lima and Dill 1990), it has been investigated directly in only a handful of primate taxa (e.g., desert baboons: Cowlishaw 1997; brown capuchins: van Schaik and van Noordwijk 1989; wedge-capped capuchins: de Ruiter 1986, Robinson 1981; see also other chapters in this volume).

By virtue of their large body size, ateline primates – woolly monkeys, spider monkeys, howler monkeys, and muriquis – are less susceptible to predation than other, smaller neotropical primates, and thus predation risk is presumed to be a less important factor influencing sociality and subsistence behavior in these species. Some researchers, in fact, have argued that the relaxation of predation pressure is what permitted the evolution of a flexible, fission–fusion pattern of social organization in large-bodied spider monkeys (Symington 1987, Terborgh and Janson 1986). While it is true that the risk of predation is likely to be much lower for atelines than for smaller-bodied platyrrhines, there is no reason to assume that ateline behaviors would not be *sensitive* to this risk. None the less, although many studies have examined the influence of ecological variables such as the abundance and distribution of foods on ateline ranging patterns, foraging strategies, and social behavior (Chapman 1988, Di Fiore and Rodman 2001, Peres 1994, Stevenson *et al.* 1994, Strier 1987, 1989, Symington 1988), few have investigated whether and how ateline behavior is shaped by the risk of predation (but see Symington 1987 for one exception).

In this chapter, I first review the available evidence implicating predation as a source of mortality for ateline primates and then discuss the limited (and largely anecdotal) body of evidence on how atelines have been observed to respond behaviorally to predators and to the threat of predation. Since reports from the literature are sparse, some of this review consists of anecdotes assembled from informal interviews of both primate researchers and indigenous hunters in western Amazonia. I then test some general predictions about how the foraging behavior of large-bodied, arboreal primates might be expected to respond to changing conditions of predation risk using data from my own studies of lowland woolly monkeys (*Lagothrix lagotricha poeppigii*). As noted above, most field studies of atelines have not focused on predation, partly because of the assumption that predation plays a secondary role to other ecological factors in influencing the behavior and social organization of these species. Thus, as an important caveat, I stress that the data used here to test various hypotheses were not collected specifically for these tests, and in some cases are not ideal for the purpose.

Predation on atelines

In 1987, Cheney and Wrangham offered the first comprehensive review of predation rates on primates. At that time, the total number of cases of predation on atelines in their data set was limited to four *suspected* incidents over a period of 7 years for one population of red howler monkeys in Venezuela. Since then, researchers have documented a number of cases of successful predation on atelines (Table 15.1).

Based on these published observations, a couple of general points can be made. First, predation on atelines – whether observed or inferred – is extremely rare. This point is underscored by my own experiences conducting research on atelines in eastern Ecuador,

Table 15.1. *Observed and suspected cases of predation on ateline primates*

Common name	Predator	Age–sex class of victim(s)	Size estimate	Description of case
Black spider monkey (*Ateles paniscus*)	Crested eagle	Juvenile, sex unspecified	2–3 kg	Directly observed predation event; predator first seen very high in sky above gap; then disappeared for 15 min before attack occurred
Black-faced spider monkey (*Ateles chamek*)	Jaguar	Unspecified	9.0 kg	Remains of victim found in scat
Black-faced spider monkey (*Ateles chamek*)	Jaguar or puma			Remains of victim found in scat
Red-bellied spider monkey (*Ateles geoffroyi*)	Puma	Unspecified	7.8 kg	Remains of victim found in scat
Red howler monkey (*Alouatta seniculus*)	Harpy eagle	Adult, sex unspecified	6.3 kg	Partial carcas of victim seen at nest of predator
Red howler monkey (*Alouatta seniculus*)	Subadult harpy eagle	Adult female		Directly observed predation attempt
Red howler monkey (*Alouatta seniculus*)	Harpy eagle	Adult male	6.5 kg	Directly observed predation event; victim was vocalizing at periphery of group; attack came while victim was facing in opposite direction
Red howler monkey (*Alouatta seniculus*)	Harpy eagle	Nearly adult male	7.3 kg	Carcass of freshly killed victim found draped over tree limb with predator present
Red howler monkey (*Alouatta seniculus*)	Jaguar	2 adult males; 1 adult female; 1 adult of un-dertermined sex; and 1 subadult female		Inferred predation events; victims disappeared one by one from study group; partial remains of the victims were found and a young jaguar was observed several times in the area
Mantled howler monkey (*Alouatta palliata*)	Jaguar	Subadult male		Directly observed predation event; victim was chased into an outdoor enclosure housing predator by members of own former social group; monkeys had never been seen to enter enclosure before
Mantled howler monkey (*Alouatta palliata*)	Jaguar	Unspecified	6.3 kg	Remains of victim found in scat
Mantled howler monkey (*Alouatta palliata*)	Puma	Unspecified	6.3 kg	Remains of victim found in scat
Muriqui (*Brachyteles arachnoides*)	Jaguar			

Source: 1. Julliot 1994, 2. Emmons 1987, 3. Chinchilla 1997, 4. Rettig 1978, Eason 1989, 6. Peres 1990,

Attack location and mode	Behavioral response from victim or group	Reference
Attack at edge of gap by predator coming from above canopy	Both howlers and spiders present near site of attack; howlers gathered at top of emergent tree and vocalized at predator; adult spiders vocalized and shook branches at predator but victim was left alone	1
		2
		2
		3
		4
Attack along edge of forest over old lake bed	Female target jumped into vine tangle then grouped with other animals on large branch near trunk of tree and howled at predator; adult male approached and threatened predator several times until it flew away, then rejoined group and vocalized with them for 30 minutes	5
Attack at top of canopy by predator soaring low over canopy		6
		7
Presumably nocturnal hunting of prey while in sleep trees		8
Presumably captured victim in tree		9
		3
		3
		10

7. Sherman 1991, 8. Peetz *et al.* 1992, 9. Cuarón 1997, 10. Olmos 1994.

where lowland woolly monkeys, red howler monkeys, and white-bellied spider monkeys all occur sympatrically, and by informal interviews I have conducted with indigenous hunters and other primatologists working in that region. In thousands of person-hours spent with groups of woolly monkeys and spider monkeys in my study site, no researchers have ever recorded a predator attack, although both atelines commonly give alarm calls to presumed aerial and terrestrial threats. Similarly, in interviews with local hunters, I have heard of only one case of successful predation on an ateline – an anecdote concerning a fight between a harpy eagle and an adult howler monkey which ended with the monkey dead and the eagle too exhausted to keep the observer from walking away with its kill.

Second, although predation on atelines is obviously rare, at least some classes of individuals in all ateline species are clearly *vulnerable* to predators. Successful predation has been observed or inferred on three of the four atelines – howler monkeys, spider monkeys, and muriquis – and most of these cases involved full grown, adult prey, likely weighing in excess of 6 to 7 kg (Table 15.1). Woolly monkeys are the only atelines for which no purported cases of predation have been reported in the literature, but even fully adult male woolly monkeys of at least one subspecies (*Lagothrix lagotricha poeppigii*) are within the range of body sizes of known kills, and juvenile individuals of all ateline species fall well within the typical prey size range of several neotropical raptors (Robinson 1994).

Third, all atelines – with the possible exception of muriquis – are potentially threatened by several different types of predators, both aerial and terrestrial. Harpy eagles are clearly the most significant potential aerial predators of atelines (and are a cause of considerable mortality for other platyrrhine species, including squirrel monkeys, capuchins, sakis, and tamarins: Fowler and Cope 1964, Rettig 1978, Robinson 1994, Terborgh 1983). Crested eagles and several species of hawk eagles, moreover, may pose an important threat to juvenile animals. Interestingly, felids may be even more significant predators of atelines than raptors, as suggested by the observations that jaguars and/or puma are known or thought to have preyed successfully on howler monkeys, spider monkeys, and muriquis (Cuarón 1997, Emmons 1987, Olmos 1994, Peetz *et al.* 1992). Finally, although there are no reported cases of large snakes preying on atelines, constrictors do prey on similar-sized primates in other parts of the world (Reichard 1998, Starin and Burghardt

1992) and are known predators of capuchin monkeys in Costa Rica (Chapman 1986). Constricting snakes are thus also potentially important predators of ateline primates.

Observed behavioral responses to the presence of predators

Based on the few available accounts, ateline behavioral responses to predator attacks principally involve alarm calling, conspicuous advertisement of the fact that a predator has been detected, and, in some cases, mobbing or attempting to chase off the predator (Table 15.1). For example, Julliot (1994) reported that members of a troop of red howler monkeys grouped together at the top of an emergent tree and howled in response to a crested eagle soaring overhead; these animals continued vocalizing until about 15 minutes after the eagle left the area with a young spider monkey it had captured. Adult spider monkeys responding to the same raptor vocalized and broke branches, a display typically prompted by threatening situations (van Roosmalen and Klein 1988). Eason (1989) noted a similar response by another group of red howler monkeys to a predation attempt by a harpy eagle. In this case, several females aggregated and howled, while the adult male howler approached and threatened the predator, and followed it through several different trees, until it flew off.

Interviews with indigenous hunters and observations by other primate researchers corroborate the above responses to potential predators as typical for atelines. For example, during a long term study of spider monkey foraging ecology in eastern Ecuador, Suarez (unpublished data) noted several cases in which adult monkeys gave repeated aerial predator alarm calls and shook branches at large raptors. In one such case involving a solitary adult female, after beginning to alarm call, the monkey approached the bird, prompting it to fly off. The spider monkey then proceeded to vocalize repeatedly for about 30 minutes as it continued foraging, and it started calling anew each time any large bird such as a macaw startled her. Such behavior has also been noted at the same research site for a subadult male woolly monkey who followed and chased off a medium-sized raptor through several tree crowns (Phillips, unpublished data). Similarly, one hunter I interviewed reported having observed a red howler monkey repeatedly directing threatening vocalizations at a harpy eagle perched in a nearby tree. This same hunter also reported seeing a group of male woolly monkeys coalesce around and shake branches at an ocelot climbing in a nearby tree. Conspicuous branch shaking and cooperative mobbing, often accompanied by what are presumed to be ground predator alarm vocalizations, are common responses for woolly, spider, and howler monkeys in eastern Ecuador when they encounter novel or potentially threatening terrestrial animals, such as human observers (Di Fiore, unpublished data). Symington (1987) has suggested that this cooperative mobbing response towards potential predators, particularly by male group members, is common across species of spider monkeys.

Observed behavioral responses to the risk of predation

Only one study has explicitly examined how any ateline primate responds behaviorally to the risk of predation. In her long-term study of the socioecology of black spider monkeys (*Ateles chamek*), Symington (1987) investigated how predation risk influences subgroup composition and the behavior of individuals in those subgroups. Reasoning that spider monkeys should be more willing to accept predation risk while foraging than while resting, she compared mean foraging party size each month to mean resting party size and found that resting party sizes were significantly larger, as predicted. Additionally, in an early field application of playback experiments, she found that spider monkeys spent significantly more time being vigilant after versus before playback of harpy eagle vocalizations. Moreover, in seven of her 19 playback trials, the subgroup to which harpy eagle calls were played left the area in response to the playback, and the decision over whether to stay or leave appeared to depend on the number of males in that subgroup; subgroups that stayed had signficantly more males than ones that left (Symington 1987).

Another long-term study of spider monkey foraging ecology provides interesting evidence of predator sensitive foraging in one specific context. Spider monkeys are known to come to the ground to feed on mineral-rich soils. Suarez (unpublished data) describes how his study animals would often spend up to several hours slowly circling one repeatedly-used soil feeding site – a small cave cut in the bank of a stream – peering at the ground, prior to descending to eat. After inspecting the area, the monkeys would finally enter the cave and feed, one at a time, for only up to 45 seconds at a stretch before rushing back up into the trees. Moreover, if something startled the animals while they were around the cave, they would often move off immediately and not return to the site that day.

Testing predator sensitive foraging hypotheses for primates

In testing whether and how predation risk influences the foraging behavior of a particular prey species, it is critical first to have an idea about what constitutes risk for members of that species (Cowlishaw 1997, Janson 1998). Given the varied environments in which primates live, the range of potential predators they face, and the breadth of antipredator strategies they might employ, there is little consensus as to how to best operationalize *intrinsic predation risk* for different primate taxa, nor are there clear ideas about how this risk is expected to change over time. None the less, a number of basic predictions – based mainly on intuition and extrapolation from other group-living animals – have been proposed that should apply to most species of social, diurnal primates. Janson (1998) summarizes four of these, suggesting that intrinsic predation risk should be greater (a) for individuals living in smaller versus larger social groups, (b) for individuals occupying peripheral versus central spatial positions within a social group, (c) for individuals farther from cover, and (d) for individuals closer to the ground. While this last suggestion relates specifically to the threat posed by terrestrial predators, an analogous proposition appropriate for aerial predators is that intrinsic predation risk should be greater (e) for animals occupying more exposed positions in a tree crown, such as the upper levels of the canopy or terminal versus more central branches (van Schaik and van Noordwijk 1989). Additionally, several authors have suggested that local neighbor density rather than social group size *per se* may be the important social factor determining an individual's risk of being preyed upon (Phillips 1995, Treves 1998), prompting the proposition that predation risk should greater (f) for animals whose nearest conspecific neighbors are farther away than for animals with closer neighbors. Finally, for group-living diurnal primates characterized by a moderate degree of sexual dimorphism, we might expect the intrinsic risk of predation to be greater (g) for females than for males because of their smaller body size and less developed secondary sexual characteristics.

Thus, one fundamental way to test for predation risk/subsistence behavior trade-offs is to examine differences in the behavior of animals found in hypothesized higher versus lower risk social situations (e.g., small versus large group membership) or occupying higher versus lower risk spatial positions (e.g., exposed versus non-exposed portions of the tree crown). Several primatological studies have used this approach, investigating, for example, whether *per capita* vigilance is greater (and foraging time, presumably, less) for individuals living in small groups or for individuals occupying riskier spatial positions within a group. The results of these studies have been mixed. Consistent with prediction (e) outlined above, van Schaik and van Noordwijk (1989) found that individual brown capuchins (*Cebus apella*) were more vigilant when *exposed* ('at a position that allowed an unobstructed approach over at least 5 m by an avian predator') than when *not exposed* in this manner. Similarly, de Ruiter (1986) found per capita vigilance for wedge-capped capuchins (*Cebus olivaceus*) to be greater for animals living in smaller groups. Rose and Fedigan (1995), in contrast, found no relationship between group size and individual vigilance for white-faced capuchins. Though this approach to studying sensitivity to predation risk is promising, it is difficult to apply *post hoc* to ateline primates, primarily because of a lack of available data. Testing the first of the seven predictions outlined above, for example, would require comparing data from several different sized groups (living in similar environments with similar predator densities), and I am aware of no such data published for any ateline primate. The remaining predictions require comparing activity data – specifically data on foraging behavior – for animals occupying different spatial positions; again, such data for atelines are largely absent from the literature. However, in my study of lowland woolly monkeys, I have collected some data of this kind and can thus attempt crude tests of several of the predator sensitive foraging predictions with this approach, comparing the behavior of animals in hypothesized high risk versus low risk spatial positions.

An alternative way to investigate the issue of predator sensitive foraging is first to examine how the risk of predation itself changes over time and then to look for coincident temporal changes in foraging behavior. Many animals are capable of assessing temporal changes in predation risk over relatively short time scales and of responding to these changes through a variety of behavioral ways (Anholt and Werner 1995, Brown 1988, Lima and Dill 1990, Nonacs and Dill 1990, Rothley *et al.* 1997, Scrimgeour and Culp 1994, Sih 1980, Suhonen 1993, Sweitzer 1996, Turner 1997). All else being equal, temporal changes in intrinsic predation risk are largely dictated by the population dynamics of both predators and their alternative prey, but good ecological information about primate predators (much less their alternative prey) is scarce (Boinski and Chapman 1995). None the less, we can still indirectly address the question of how temporal changes in predation risk affect foraging behavior by recognizing that all else is *not* equal and that 'predation risk' is just one component of the 'general risk' to which animals should be sensitive. Many classic foraging experiments show that animals are willing to expose themselves to conditions of greater predation risk when food resources are scarce – that is, when the 'risk of inadequate food acquisition' is high. If this is true for large-bodied primates as well, then the propositions outlined above can be tested by examining how an animal's decisions over *exposure to predation risk* – reflected in its choices over where and how to forage – change with resource availability in a longitudinal way.

Rephrasing six of the seven predictions outlined above, we would expect animals to spend more time in high risk areas – in peripheral spatial positions, farther from cover, closer to the ground, in the upper portions of the canopy, and in more exposed positions in tree crowns – when food is scarce; we also would predict animals to be farther from other individuals or to have fewer individuals around them when food is scarce. Through my research on the foraging ecology of lowland woolly monkeys, I have documented how the availability of resources important to this species changes over time (Di Fiore 1997, Di Fiore and Rodman 2001). Thus, I am able to use this second approach to test a number of predicted relationships between predation risk and foraging behavior.

Methods

Study area and subjects

Data were collected on several habituated groups of lowland woolly monkeys found at my study site in Yasuní National Park, Ecuador between August 1994 and April 1996. The roughly 500-ha study site consists largely of intact, primary tropical rain forest spread out over a series of ridges and drainages. Nine other species of nonhuman primates are found in the study area, including two other genera of large-bodied atelines – white-bellied spider monkeys (*Ateles belzebuth*) and red howler monkeys (*Alouatta seniculus*). Woolly monkeys are by far the most numerous and most frequently encountered species of primate in the site, and the biomass of woolly monkeys per hectare in Yasuní is the highest yet reported. Behavioral data for this study come primarily from two groups of woolly monkeys that ranged closest to a narrow gravel road providing access to the western part of Yasuní National Park. As elsewhere in their geographic range, woolly monkeys in Yasuní live in large multimale–multifemale groups typically containing between 15 and 30 individuals (Defler and Defler 1995, Di Fiore and Rodman 2001, Kinzey 1997, Nishimura 1990, Ramirez 1988). The groups that are the focus of this study each contained around 24 indviduals.

The potential predator community in and around the study site is largely intact. Observers regularly encounter tracks of large felids (jaguar and puma) on trails in the study site, and large raptors (harpy eagles, crested eagles, several species of hawk eagles, and large owls) have been seen either in the study site or elsewhere along the gravel road accessing the park. Similarly, within the study site, the community of alternative prey species used by potential ateline predators is also intact. Smaller ceboid primates, peccaries, deer, kinkajous, sloths, and many other species of small mammals and birds are very abundant.

Behavioral sampling

Between April 1995 and March 1996, I spent between 16 and 20 days each month conducting full-day follows of these two groups with occasional follows of additional social groups. Each group was followed for two 5-day periods each month. During a 5-day follow, I used a combination of instantaneous scan sampling and focal animal sampling (Altmann 1974) to collect data on behavior and microhabitat use, with scan and focal samples collected on alternate days. On scan days, I began sampling 5 minutes after a group was first encountered and took scans at 10-minute intervals thereafter for the remainder of the day. A scan lasted 5 minutes and was followed by 5 minutes of inactivity until the next scan began. Efforts were made to collect data on as many different individuals as possible during a scan by changing my position under the group frequently. However, since group members were often spread out over a large area, typically only a few individuals were able to be sampled during any given scan (mean $= 4.76$, SD $= 1.79$, range $= 1$–15, $n = 4180$).

During each scan, the age–sex class, behavior, position in the tree crown, height in the canopy, and nearest neighbors of each individual woolly monkey that came into view over the course of the scan period were recorded. I refer to this set of data pertaining to each scanned animal as an individual *behavioral record*, and a total of 19 519 behavioral records were collected over the course of the study. Behaviors were divided into seven mutually exclusive categories: moving, resting, feeding on fruits, foraging for insects (including scanning for animal prey), socializing, vigilance scanning, and other. Vigilance scanning, defined as actively scanning the social or biotic environment from a stationary position, was a rare behavior, distinguishable from much more common 'prey scanning' which was always clearly directed at substrates close to an animal's body. During a scan, the position of each scanned animal within a tree crown was recorded, and animals occupying positions in the terminal branches, on below-crown lianas, and on the trunks of trees were considered to be more *exposed* to predators than animals occupying positions in the central tree crown or in tangles of lianas. Height in the canopy was divided into 5 m strata during data collection and grouped *post hoc* into three categories: lower strata (0–15 m), middle strata (15–25 m), and upper strata (25+ m). Additionally, I marked all trees fed in by a group for more than 5 minutes and either estimated or measured the heights of these feeding trees to the nearest meter.

For characterizing activity and substrate use patterns, I used each behavioral record as a separate data point. This approach is appropriate for examining decisions over substrate use given the distance over which woolly monkeys typically spread out while foraging (Di Fiore and Rodman 2001, Peres 1996) and the range of substrates thus available to them. This method is more problematic, however, when used to characterize activity patterns since it has the potential effect of biasing estimates of time allocation towards conspicuous, aggregate behaviors – such as group feeding in large fruiting trees – when variable numbers of individuals are observed per scan (Clutton-Brock 1977). In this study, however, such a bias is likely to be minimal for several reasons. In most cases, even foraging activity within a tree crown was not highly synchronized among group members. Moreover, I have demonstrated elsewhere that time budgets for this population of woolly monkeys calculated using individual behavioral records are essentially equivalent to those derived by first calculating separate budgets for each scan and then averaging across the set of scan budgets (Di Fiore 1997). This suggests that nonindependence of activities between individuals does not present a serious problem for estimating time allocation variables (e.g., the proportion of time devoted each month to fruit feeding or to foraging for animal prey) which are then used as single cases in statistical analyses. However, using each behavioral record as a separate data point could violate the notion of independent observations and create the statistical problem of artificially inflating counts of rare behaviors if there is a social component to these behaviors. Vigilance could pose such a problem if, for

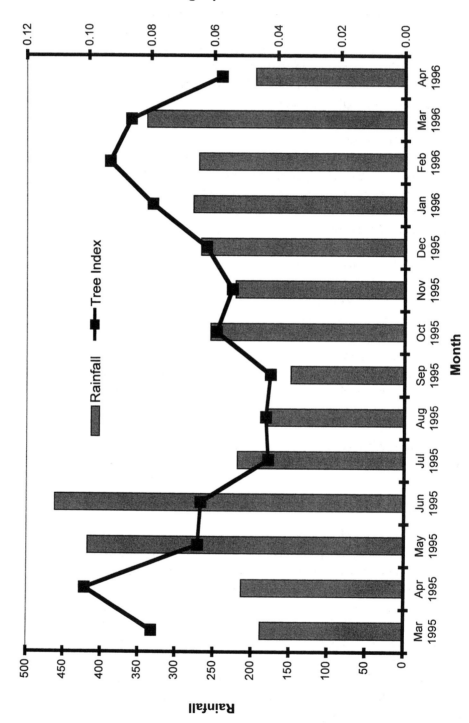

Fig. 15.1. Monthly changes in rainfall and in the habitat-wide availability of ripe fruits.

example, woolly monkeys are more likely to be vigilant when conspecifics are vigilant. Although vigilance overall was rare in this study, over 10% of the behavioral records in which vigilance was recorded did occur during scans in which more than one animal was vigilant simultaneously. Thus, in order not to inflate the sample size for the occurrence of this behavior artificially, I used only one randomly selected record of vigilance per scan in the microhabitat use and vigilance analyses below.

Phenological sampling

To provide an independent estimate the abundance of plant resources available to the woolly monkey population, I established five 1-ha belt transects at random locations relative the trail system in the site. Each transect was divided into 100, 10 m × 10 m blocks, half of which were then selected for phenological sampling. The total 2.5 ha of phenology blocks contained 1492 trees with a diameter at breast height (DBH) of greater than or equal to 10 cm. Each month from March 1995 to April 1996, a field assistant recorded the presence and abundance of new leaves or leaf flush, flowers, immature fruits and ripe fruits present in each tree crown and in any associated epiphytic plants. From these data, I constructed several indices of the monthly abundance of particular plant phenophases (e.g., new leaves, flowers, immature fruits, ripe fruits) (Di Fiore and Rodman 2001). Because ripe fruit is the predominant component in the woolly monkey diet, constituting an average of 77% of the food items consumed each month (Di Fiore and Rodman 2001), I use here one of the indices of ripe fruit abundance as a measure of the monthly availability of food to the woolly monkeys (Fig. 15.1).

Specific predictions and results

Microhabitat use and vigilance

If woolly monkey foraging is sensitive to the risk of predation, we would expect them to perform behaviors associated with predator detection more often when in vulnerable microhabitats. Vigilance activity is considered one such behavior, and vigilant animals are presumed to be more likely to detect predators than nonvigilant animals. Even where vigilance is directed primarily at conspecifics, an animal's risk of being preyed upon is likely to decrease with increasing time spent vigilant either because they are personally more likely to detect a predator, or because they are likely to be alerted more quickly to the presence of a predator by cueing in on the behavior of others. Thus, we would predict woolly monkeys to be more vigilant while in the upper portion of the canopy, where the risk from avian predators is greatest, and while near the ground, where the risk from terrestrial predators is greatest. Similarly, we would predict woolly monkeys to be more vigilant while on exposed substrates.

Examining the distribution of vigilance activity by height in the canopy reveals a significant deviation from that expected by chance ($\chi^2 = 18.29$, df = 2, $p < 0.001$). Woolly monkeys are more vigilant than expected when close to the ground, but are less vigilant than expected when in upper levels of the canopy where *a priori* we might have predicted they would be most susceptible to predation.

Examining the distribution of vigilance activity by exposure (terminal branches, below and between crown lianas, and trunks versus central crowns and vine tangles) again reveals significant deviation from that expected by chance ($\chi^2 = 45.66$, df = 1, $p < 0.001$). Contrary to predicted, however, animals were much less vigilant than expected when on *more exposed* substrates, where they are presumed to be most at risk from predators, and more vigilant while in other, safer portions of the tree crown.

Microhabitat use and food availability

If more exposed areas are associated with higher predation risk, and if animals are willing to accept higher levels of predation risk while foraging during times of resource scarcity, then we would expect to see a negative association between food availability and time spent in the upper and lower strata of the forest where the risk of arboreal and terrestrial predation, respectively, are presumed to be greatest. In testing for such associations, however, it is necessary to control for the fact that fruits are presented in particular portions of the canopy and monkeys have to occupy those parts of the tree crown in order to feed. I thus examined the partial correlations between the amount of time woolly monkeys spent in lower (below 15 m) and upper (above 25 m) strata of the forest and the availability of ripe fruits while controlling for the time woolly monkeys spent feeding on fruits each month and for the mean height of feeding trees used each month.

Contrary to the prediction, the proportion of behavioral records in which woolly monkeys were recorded in lower strata of the forest each month was not correlated with the monthly availability of ripe fruit (partial correlation $r = 0.170$, $n = 12$, NS (not significant), controlling for the proportion of time spent eating fruits each month and the mean height of feeding trees). Moreover, the use of upper strata (25+ m) was significantly positively associated with fruit availability rather than negatively as predicted (partial correlation $r = 0.620$, $n = 12$, $p < 0.05$).

The above tests examined *overall time* spent in particular levels of the forest as an index of the degree to which animals are willing to accept predation risk. However, it may be more appropriate to consider simply differential strata use for *subsistence activity* only – time spent feeding on fruits plus foraging for insects. Again, there was no significant relationship between the use of lower strata and ripe fruit availability (partial $r = 0.160$, $n = 12$, NS, controlling for proportion of time spent eating fruits per month and mean feeding source height per month). Use of upper strata for subsistence activity was again greater during periods of higher rather than lower ripe fruit availability, and this relationship approached significance (partial $r = 0.587$, $n = 12$, $p < 0.10$). Thus, contrary to the predator sensitive foraging prediction, woolly monkeys used forest strata associated with greater hypothesized predation risk *less* often rather than more often when resources were scarce.

We might also predict that when fruit is least abundant, animals would be willing to accept higher predation risk by being on more exposed substrates, such as at the periphery of the tree crown or under the canopy layer. There was no significant relationship between the total amount of time in which animals were recorded in the exposed portions of a tree and the availability of ripe fruits that month, and the correlation coefficient is in fact in the opposite direction to that predicted (partial $r = 0.292$, $n = 10$, NS, controlling for proportion of time spent eating fruits per month and mean feeding tree height). If we just consider time spent in subsistence activity in exposed positions (rather than the total time spent in exposed positions), then we do find a negative relationship between exposure and monthly ripe fruit availability, though the relationship is not significant (partial $r = -0.414$, $n = 10$, NS). Thus, woolly monkeys spent more rather than less time in 'risky' positions when fruits were more available but a smaller proportion of that time was spent on subsistence activity, counter to what we would predict for predator sensitive foragers.

Interindividual distance and food availability

If predation risk is an important determinant of woolly monkey foraging behavior and if resource scarcity forces animals more readily to expose themselves to the risk of predation, we would predict a clear association between resource availability and proximity of neighbors. Specifically, we would expect animals to have

Fig. 15.2. Relationship between fruit availability and foraging time with a neighbor in close proximity.

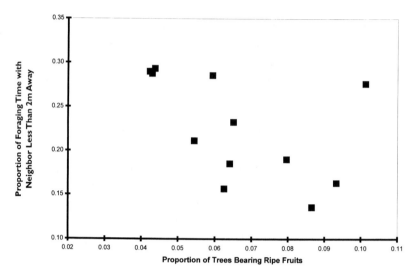

Fig. 15.2. Relationship between fruit availability and foraging time with a neighbor in close proximity.

neighbors around them less often when ripe fruit is scarce. Note that this same relationship would be predicted if *intragroup feeding competition* over ripe fruits, rather than predation risk, were shaping patterns of spatial association between individuals; that is, we would predict that animals should spread out when food is least abundant to minimize competition with groupmates over food.

Contrary to prediction, the proportion of time that woolly monkeys had a nearest neighbor within 10 m of them was not related to the availability of ripe fruits (Spearman rank correlation: $r_s = 0.133$, $n = 12$, NS), nor was the proportion of time that an animal had neighbors in close proximity (defined as within 2 m) related to ripe fruit availability ($r_s = -0.343$, $n = 12$, NS). Furthermore, considering only those associations while animals are engaged in subsistence activity rather than nearest neighbor spatial associations as a whole, there is still no relationship between ripe fruit availability each month and the proportion of time an animal had a neighbor within 10 m (Spearman rank correlation: $r_s = -0.098$, $n = 12$, NS). The proportion of time an animal foraged with a nearest neighbor within 2 m, however, was strongly *negatively* related to food availability (Spearman rank correlation: $r_s = -0.636$, $n = 12$, $p < 0.05$). Thus, animals spaced out *more* while foraging when fruit was abundant rather than *less* as a predator sensitive foraging hypothesis – or a hypothesis concerning *intragroup feeding competition* over fruit resources – would suggest. This observation is further underscored by the fact that the mean distance from an animal to its nearest neighbor was positively correlated with fruit availability (Spearman rank correlation: $r_s = 0.685$, $n = 12$, $p < 0.05$) rather than negatively as would be predicted (Fig. 15.2). It might be argued that close associations between mothers and juveniles could influence these results; however, repeating all of these analyses considering only nonjuvenile animals and their nonjuvenile nearest neighbors yields an identical pattern.

Interindividual distance and microhabitat use

If animals are more at risk from predators while on exposed substrates, and if the presence of neighbors provides some protection against predators, either through dilution or enhanced predator detection effects, then we would predict that animals using exposed substrates should stay closer to neighbors than animals on less exposed substrates. I tested this hypothesis by comparing the proportion of time animals had neighbors either within 10 m or within 2 m to their use of more exposed versus less exposed substrates.

Considering first the set of all animals, both nonjuveniles and juveniles, individuals in *more exposed* positions had nearest neighbors within 10 m of them more often than expected by chance (χ^2 = 18.85, df=1, $p<0.001$), but there was no significant deviation from expected in the amount of time neighbors were within 2 m ($\chi^2=1.11$, df=1, NS). If only nonjuvenile animals are considered, there was no significant deviation from expected in the amount of time animals in *more exposed* positions had nearest neighbors within 10 m of them ($\chi^2=1.60$, df=1, NS), but nearest neighbors within 2 m were more common than expected ($\chi^2=10.68$, df=1, $p<0.01$). If we consider foraging time only and look at the set of all individuals, *more exposed* animals had neighbors both within 10 m of them and within 2 m of them more often than expected by chance (within 10 m: $\chi^2=35.30$, df=1, $p<0.001$; within 2 m: $\chi^2=9.15$, df=1, $p<0.01$). These results are the same if only nonjuvenile animals and their nonjuvenile neighbors are considered (within 10 m: $\chi^2=12.36$, df=1, $p<0.001$; within 2 m: $\chi^2=5.90$, df=1, $p<0.05$). Taken together, these results suggest that woolly monkeys on *more exposed* substrates were significantly more likely to have closest neighbors within 10 m or within 2 m of them than expected by chance, both while foraging and overall. These results are consistent with the idea that woolly monkeys are sensitive to an increased risk of predation while in exposed positions, but this sensitivity may not be limited to the foraging context.

If upper and lower strata in the forest are associated with greater vulnerability to predation, then woolly monkeys should also have neighbors nearby more often when using these strata. Across forest strata, the proportion of time woolly monkeys had neighbors within 10 m did deviate significantly from expected ($\chi^2 = 77.97$, df = 4, $p < 0.001$); animals had nearest neighbors within 10 m more often than expected while in upper and middle strata of the forest, but less often than expected while in lower strata. However, this result was only seen if juveniles and juvenile neighbors were included in the analysis. Considering only nonjuveniles and their nonjuvenile neighbors, there was no significant deviation from expected in the proportion of time that animals had neighbors within 10 m by canopy level ($\chi^2 = 8.44$, df = 4, NS). Moreover, woolly monkeys had neighbors within 2 m significantly less often than expected by chance while in upper and lower forest strata whether all animals ($\chi^2 = 155.62$, df = 4, $p < 0.001$) or nonjuveniles only ($\chi^2 = 31.71$, df = 4, $p < 0.001$) are considered, rather than more often as was predicted. Thus, the spatial relationships of woolly monkeys using different forest strata do not conform to the pattern predicted by predator sensitive foraging hypotheses.

Sex differences

One final way to examine woolly monkey behavioral responses to the risk of predation is to look at differences between vigilance activities and microhabitat use decisions of males and females. If males do face decreased predation risk because of their larger body size and greater development of defensive physical attributes such as canines, we would expect male woolly monkeys to use upper and lower strata of the forest and exposed positions relatively more often than females and to do more of their foraging in these areas of the forest. We would also expect males to be less vigilant than females and to have neighbors close by them less often.

There was a significant overall difference between nonjuvenile male and female woolly monkeys in use of forest strata ($\chi^2 = 32.94$, df = 4, $p < 0.001$), with males using lower forest strata more often than expected and females using lower levels less often then expected. Use of upper forest strata, however, did not follow the predicted pattern. Males and females also differed significantly in their use of more exposed versus less exposed substrates ($\chi^2 = 16.94$, df = 1, $p < 0.001$) but, counter to prediction, females used exposed substrates slightly more often than males (61% versus 57% of behavioral records). Males and females also differed significantly in vigilance behavior, but, counter to prediction, males were more vigilant than expected while females were less vigilant than expected ($\chi^2 = 5.38$, df = 1, $p < 0.05$). Finally, we might expect females with dependents to be more sensitive to the risk of predation and thus to be more vigilant than females without dependents. In fact, the reverse was true;

females with juveniles spent less time being vigilant than expected, while females without juveniles were more vigilant than expected ($\chi^2 = 5.48$, df = 1, $p < 0.05$), in spite of the fact that females without juveniles also engaged in significantly more subsistence behavior (42% versus 32% of their respective activity budgets: Di Fiore 1997).

Discussion

Taken together, these results suggest that woolly monkeys do little to adjust their foraging behavior to the intrinsic risk of predation, at least not in the simple ways we might expect. While woolly monkeys are more vigilant than expected when close to the ground, they are not more vigilant while occupying other hypothesized vulnerable microhabitats (upper levels of the canopy or more exposed substrates); this result is consistent with the idea that woolly monkeys may perceive terrestrial predators (such as large felids) as posing a more significant threat than arboreal predators (such as raptors). Furthermore, woolly monkeys do not use hypothesized vulnerable microhabitats more when food resources are scarce, nor do they have neighbors around them more often or more close by when food is abundant. Moreover, although there is some evidence that they do maintain closer distances to neighbors while using hypothesized vulnerable substrates (but not while using presumably more vulnerable canopy levels), this does not appear to be a response specific to the foraging context. Finally, females are not more vigilant than males and do not spend less time than males in hypothesized vulnerable microhabitats. Thus, it seems clear that the foraging behavior of woolly monkeys is largely insensitive to sex differences in intrinsic predation risk or to temporal changes in predation risk on the scale examined here.

In contrast, the finding that woolly monkeys are more vigilant in less exposed versus more exposed microhabitats could be interpreted as indicative of sensitivity to predation risk as it does accord with an alternative hypothesis as to how vigilance might be employed in an antipredation context – a hypothesis we might refer to as the 'scan from cover' model. If predators are less able to attack animals successfully in 'covered' microhabitats, then we might expect animals to use those microhabitats as sites from which to scan for predators before venturing out into more vulnerable areas. This model would then predict a positive association between vigilance and use of less exposed microhabitats (as seen here) and perhaps a negative association between vigilance and distance from a refuge (counter to results from desert baboons demonstrating increased vigilance with distance from a refuge: Cowlishaw 1997). Further exploration of this idea would require more fine-scaled analyses of the temporal association between vigilance behavior and microhabitat use. For example, do animals enter vulnerable micro-habitats immediately after scanning the environment?

Additionally, if it is true that female woolly monkeys face greater intrinsic risk of predation than males (which is clearly an assumption open to debate), then the finding that females spent more time than males in hypothesized vulnerable microhabitats may be consistent with yet another model of the relationship between vigilance and predator detection. If predators are detected more easily by animals in more exposed microhabitats (perhaps because their vision is less obstructed by intervening substrates), or if predators can be detected at greater distances, then we might expect animals to spend more time in these microhabitats than in less exposed ones. Under this 'better detection in the open' model, there is no clear prediciton about relative levels of vigilance we would expect to see employed in more versus less exposed microhabitats, since detection success would be a function of both vigilance time and microhabitat features. Thus, observations of greater vigilance by animals using less exposed substrates (as seen in this study) would not be inconsistent with this model. This is an issue that deserves much more attention in future studies of predator sensitive behavior in primates.

While I have only provided tests of predator sensitive foraging hypotheses for woolly monkeys, I suspect that these results will hold true for most other ateline primates as well. These results do not necessarily suggest that predation risk is not an important determinant of sociality in ateline primates, nor that the rate of predation on atelines is insignificant (although it would take much to demonstrate convincingly either of these points). Clearly, at least some classes of individuals within each ateline genus *are* susceptible to predation, potentially from several different types of predators, including raptors (such as crested and harpy eagles), large-bodied felids (such as jaguar and puma), and perhaps constricting snakes (such as boas). None the less, I would argue that for woolly monkeys, patterns of microhabitat use and nearest neighbor relationships are motivated far more by the search for food than by trying to minimize predation risk. In fact, the results presented here support the idea that when fruit resources are plentiful, woolly monkeys follow a foraging strategy for maximizing food acquisition – specifically, by devoting additional time to foraging for animal prey – rather than one that allows them to keep safely away from predators (Di Fiore and Rodman 2001).

One of the general problems with studying the impact of predation and predation risk on the behavior and social systems of animals is that predator–prey relationships are complex and poorly understood. For the specific issue of predator sensitive foraging, the major difficulty associated with examining trade-offs between predation risk and subsistence activity is that there are few generalities as to how foraging behavior *should* respond to changing predation risk. In part, this is because the responses we might expect depend fundamentally on various aspects of the natural history of the prey species, including its foraging strategy, its social structure, and way in which resources important to its survival are distributed in space and vary over time. Additionally, different kinds of predators impose very different selective pressures; therefore, identifying how – or even if – a particular primate responds to changing predation risks depends on accurately determining what those risks are and who poses them. As an example, the behavioral strategy that spider monkeys employ to reduce the risk of predation from diurnal raptors would undoubtedly be very different from that used against crepuscular, terrestrial felids. We might expect predator sensitive spider monkeys to minimize the amount of time they spend foraging in the upper branches of a tree crown or in other exposed portions of the canopy as a hedge against the former, while perhaps minimizing the time they spend on the ground foraging for mineral-rich soils as protection against the latter.

Finally, in order to assess truly the influence of predators and predation risk on primates, we need to have a far better understanding of the predators themselves (Boinksi and Chapman 1995), an important point that deserves reiterating. How does predator behavior (ranging patterns, hunting strategy, and so on) change seasonally? How does predator density change over time and space? What are the population dynamics of the predator independent of the primate prey under consideration? What alternative prey types are there in the predator's diet, what are their densities and population dynamics, and how do they themselves react to temporal changes in predation risk? Also, how do local environmental conditions (like the location and density of refuges) or meteorological conditions and cycles (like daily temperature fluctuations or cloud cover or phases of the moon) influence a predator's hunting strategy? Unfortunately, we still know very little about the basic behavior and ecology of many primate predators – and less still about the nuances of their foraging strategies and population cycles and how these, in turn, depend on the ecologies of alternative prey species. It is time for primatologists interested in predation and its effects to take a more explicit community-level perspective on these issues as they are critical to addressing the issue of predator sensitive behavior in primates.

Acknowledgments

I am very grateful to the government of Ecuador and to the officials of INEFAN (Instituto Ecuatoriano Forestal y de Areas Naturales y de Vida Silvestre) and the Ministerio de Ambiente for permission to conduct this research, and for their continuing interest in primate research and conservation in eastern Ecuador. Dr. Luis Albuja of the Escuela Politécnica Nacional of Quito served as my counterpart in Ecuador for this research and deserves thanks for help and friendship he provided throughout the study. Dr. Laura Arcos Terán and Dr. Alberto Padilla of the Pontifícia Universidad Católica de Ecuador provided fundamental logistical support, and Maxus Ecuador, Inc. generously contributed additional logistical assistance to all phases of this study. Very special thanks are due to my partner, Dr. Kristin Phillips, who provided invaluable help and companionship throughout the field work, and to Dr. Peter Rodman who was instrumental in initiating this research in Ecuador and is a much respected mentor, colleague, and friend. I am also especially grateful to Michelle Avallone with whom I had many thoughtful discussions concerning predation and predator sensitive foraging. This manuscript benefited from the thoughtful comments of Peter Rodman, Lynne Miller, Mark Prescott, and an anonymous reviewer. Funding for this study was provided by grants from the University of California, Davis, the National Science Foundation, the LSB Leakey Foundation, and the Wenner–Gren Foundation for Anthropological Research.

REFERENCES

Alexander, R.D. (1974). The evolution of social behavior. *Annual Review of Ecology and Systematics* **5**: 325–83.

Altmann, J. (1974). Observational study of behavior: sampling methods. *Behaviour* **49**: 227–62.

Anholt, B.R., and Werner, E.E. (1995). Interaction between food availability and predation mortality mediated by adaptive behavior. *Ecology* **76**: 2230–4.

Boinski, S., and Chapman, C.A. (1995). Predation on primates: where are we and what's next? *Evolutionary Anthropology* **4**: 1–3.

Brown, J.S. (1988). Patch use as an indicator of habitat preference, predation risk, and competition. *Behavioral Ecology and Sociobiology* **22**: 37–47.

Chapman, C.A. (1986). Boa constrictor predation and group response in white-faced cebus monkeys. *Biotropica* **18**: 171–2.

Chapman, C. (1988). Patch use and patch depletion by the spider and howling monkeys of Santa Rosa National Park, Costa Rica. *Behaviour* **105**: 99–116.

Cheney, D.L., and Wrangham, R.W. (1987). Predation. In: B.B. Smuts, D.L. Cheney, R.M. Seyfarth, R.W. Wrangham, and T.T. Struhsaker, eds., *Primate Societies*, Chicago: University of Chicago Press, pp. 227–39.

Chinchilla, F.A. (1997). La dieta del jaguar (*Panthera onca*), el puma (*Felis concolor*), y el manigordo (*Felis pardalis*) (Carnivora: Felidae) en el Parque Nacional Corcovado, Costa Rica. *Revista de Biología Tropicál* **45**: 1223–9.

Clutton-Brock, T.H. (1977). Appendix I: methodology and measurement. In: T.H. Clutton-Brock, ed., *Primate Ecology: studies of foraging and ranging behaviour in lemurs, monkeys, and apes*, London: Academic Press, pp. 585–90.

Cowlishaw, G. (1997). Trade-offs between foraging and predation risk determine habitat use in a desert baboon population. *Animal Behaviour* **53**: 667–86.

Cuarón, A.D. (1997). Conspecific aggression and predation: costs for a solitary mantled howler monkey. *Folia Primatologica* **68**: 100–5.

Defler, T.R., and Defler, S.B. (1995). Diet of a group of *Lagothrix lagotricha lagotricha* in Southeastern Colombia. *International Journal of Primatology* **17**: 161–90.

de Ruiter, J.R. (1986). The influence of group size on predator scanning and foraging behaviour of wedge-capped capuchin monkeys (*Cebus olivaceus*). *Behaviour* **98**: 240–58.

Di Fiore, A. (1997). Ecology and Behavior of Lowland Woolly Monkeys (*Lagothrix lagotricha poeppigii*, Atelinae) in Eastern Ecuador. Ph.D. Dissertation, University of California, Davis.

Di Fiore, A., and Rodman, P.S. (2001). Time allocation patterns of lowland woolly monkeys (*Lagothrix lagotricha poeppigii*) in a neotropical *terra firma* forest. *International Journal of Primatology* **22**: 449–80.

Dunbar, R.I.M. (1988). *Primate Social Systems*. Ithaca: Cornell University Press.

Eason, P. (1989). Harpy eagle attempts predation on adult howler monkey. *The Condor* **91**: 469–70.

Emmons, L.H. (1987). Comparative feeding ecology of felids in a neotropical rainforest. *Behavioral Ecology and Sociobiology* **20**: 271–83.

Fowler, J.M., and Cope, J.B. (1964). Notes on the harpy eagle in British Guiana. *The Auk* **81**: 257–73.

Fraser, D.F., and Huntingford, F.A. (1986). Feeding and avoiding predation hazard: The behavioural response of the prey. *Ethology* **73**: 56–68.

Isbell, L.A. (1994). Predation on primates: ecological patterns and evolutionary consequences. *Evolutionary Anthropology* **3**: 61–71.

Janson, C.H. (1998). Testing the predation hypothesis for vertebrate sociality: prospects and pitfalls. *Behaviour* **135**: 389–410.

Julliot, C. (1994). Predation of a young spider monkey (*Ateles paniscus*) by a crested eagle (*Morphnus guianensis*). *Folia Primatologica* **63**: 75-7.

Kinzey, W.G. (ed.) (1997). *New World Primates: Ecology, Evolution, and Behavior*. New York: Aldine de Gruyter.

Lima, S.L., and Dill, L.M. (1990). Behavioral decisions made under the risk of predation: a review and prospectus. *Canadian Journal of Zoology* **68**: 619-40.

Nishimura, A. (1990). A sociological and behavioral study of woolly monkeys, *Lagothrix lagotricha*, in the Upper Amazon. *Science and Engineering Review of Doshisha University* **31**: 87-121.

Nonacs, P., and Dill, L.M. (1990). Mortality risk vs. food quality trade-offs in a common currency: ant patch preferences. *Ecology* **71**: 1886-92.

Olmos, F. (1994). Jaguar predation on muriqui, *Brachyteles arachnoides*. *Neotropical Primates* **2**: 16.

Peetz, A., Norconk, M.A., and Kinzey, W.G. (1992). Predation by jaguar on howler monkeys (*Alouatta seniculus*) in Venezuela. *American Journal of Primatology* **28**: 223-8.

Peres, C.A. (1990). A harpy eagle successfully captures an adult male red howler monkey. *Wilson Bulletin* **102**: 560-1.

Peres, C.A. (1994). Diet and feeding ecology of gray woolly monkeys (*Lagothrix lagotricha cana*) in Central Amazonia: comparisons with other atelines. *International Journal of Primatology* **15**: 333-72.

Peres, C.A. (1996). Use of space, spatial group structure, and foraging group size of gray woolly monkeys (*Lagothrix lagotricha cana*) at Urucu, Brazil. In: M.A. Norconk, A.L. Rosenberger, P.A. Garber, eds., *Adaptive Radiations of Neotropical Primates*. New York: Plenum Press, pp. 467-88.

Phillips, K.A. (1995). Resource patch size and flexible foraging in white-faced capuchins (*Cebus capuchinus*). *International Journal of Primatology* **16**: 509-19.

Ramirez, M. (1988). The woolly monkeys, genus *Lagothrix*. In: R.A. Mittermeier, A.B. Rylands, A.F. Coimbra-Filho, and G.A.B. da Fonseca, eds., *Ecology and Behavior of Neotropical Primates*, Vol. 2. Washington, DC: World Wildlife Fund, pp. 539-75.

Reichard, U. (1998). Sleeping sites, sleeping places, and presleep behavior of gibbons (*Hylobates lar*). *American Journal of Primatology* **46**: 35-62.

Rettig, N.L. (1978). Breeding behavior of the harpy eagle (*Harpia harpyja*). *The Auk* **95**: 629-43.

Robinson, J.G. (1981). Spatial structure in foraging groups of wedge-capped capuchin monkeys (*Cebus nigrivittatus*). *Animal Behaviour* **29**: 1036-56.

Robinson, S.K. (1994). Habitat selection and foraging ecology of raptors in Amazonian Peru. *Biotropica* **26**: 443-58.

Rose, L.M., and Fedigan, L.M. (1995). Vigilance in white-faced capuchins, *Cebus capuchinus*, in Costa Rica. *Animal Behaviour* **49**: 63-70.

Rothley, K.D., Schmitz, O.J., and Cohon, J.L. (1997). Foraging to balance conflicting demands: novel insights from grasshoppers under predation risk. *Behavioral Ecology* **8**: 551-9.

Scrimgeour, G.J., and Culp, J.M. (1994). Feeding while evading predators by a lotic mayfly: linking short-term foraging behaviors to long-term fitness consequences. *Oecologia* **100**: 128–34.

Sherman, P.T. (1991). Harpy eagle predation on a red howler monkey. *Folia Primatologica* **56**: 53–6.

Sih, A. (1980). Optimal behavior: can foragers balance two conflicting demands? *Nature* **210**: 1041–3.

Starin, E.D., and Burghardt, G.M. (1992). African rock pythons (*Python sebae*) in The Gambia: Observations on natural history and interactions with primates. *Snake* **24**: 50–62.

Stevenson, P.R., Quiñones, M.J., and Ahumeda, J.A. (1994). Ecological strategies of woolly monkeys (*Lagothrix lagotricha*) at Tinigua National Park, Colombia. *American Journal of Primatology* **32**: 123–40.

Strier, K.B. (1987). Ranging behavior of woolly spider monkeys, or muriquis, *Brachyteles arachnoides*. **8**: 575–91.

Strier, K.B. (1989). Effects of patch size on feeding associations in muriquis (*Brachyteles arachnoides*). *Folia Primatologica* **52**: 70–7.

Suhonen, J. (1993). Risk of predation and foraging sites of individuals in mixed species tit flocks. *Animal Behaviour* **45**: 1193–8.

Sweitzer, R.A. (1996). Predation or starvation: consequences of foraging decisions by porcupines (*Erethizon dorsatum*). *Journal of Mammalogy* **77**: 1068–77.

Symington, M.M. (1987). Ecological and Social Correlates of Party Size in the Black Spider Monkey, *Ateles paniscus chamek*. Ph.D. Dissertation, Princeton University.

Symington, M.M. (1988). Food competition and foraging party size in the black spider monkey (*Ateles paniscus chamek*). *Behaviour* **105**: 117–32.

Terborgh, J. (1983). *Five New World Primates: A Study in Comparative Ecology. Monographs in Behavior and Ecology*. Princeton: Princeton University Press.

Terborgh, J., and Janson, C.H. (1986). The socioecology of primate groups. *Annual Review of Ecology and Systematics* **17**: 111–35.

Treves, A. (1998). The influence of group size and neighbors on vigilance in two species of arboreal monkeys. *Behaviour* **135**: 453–81.

Turner, A.M. (1997). Contrasting short-term and long-term effects of predation risk on consumer habitat use and resources. *Behavioral Ecology* **8**: 120–5.

van Roosmalen, M.G.M., and Klein, L.L. (1988). The spider monkeys, genus *Ateles*. In: R.A. Mittermeier, A.B. Rylands, A.F. Coimbra-Filho and G.A.B. de Fonesca, eds., *Ecology and Behavior of Neotropical Primates*, Vol. 2, Washington, DC: World Wildlife Fund, pp. 455–537.

van Schaik, C.P. (1983). Why are diurnal primates living in groups? *Behaviour* **87**: 120–44.

van Schaik, C.P., and van Noordwijk, M.A. (1989). The special role of male *Cebus* monkeys in predation avoidance and its effect on group composition. *Behavioral Ecology and Sociobiology* **24**: 265–76.

Wright, P.C. (1998). Impact of predation risk on the behaviour of *Propithecus diadema edwardsi* in the rain forest of Madagascar. *Behaviour* **135**: 483–512.

16 • Antipredatory behavior in gibbons (*Hylobates lar*, Khao Yai/Thailand)

NICOLA L. UHDE & VOLKER SOMMER

Introduction

Gibbons are small diurnal apes inhabiting the rain forests of South Asia (Preuschoft *et al.* 1984). The social system is predominantly monogamous and groups have typically only two to six members. Recent studies revealed the existence of extra-pair copulations, partner changes and at least occasional polyandrous or polygynous group composition (Brockelman *et al.* 1998, Palombit 1994a,b, Reichard 2001, Sommer and Reichard 2000).

Primate species that form single-female groups are relatively rare (Kleiman 1977). Selection pressures such as intergroup competition for food (van Schaik 1989, Wrangham 1980) and predation risk (van Schaik 1983, van Schaik and van Hooff 1983) are believed to favor the development of (polygynous or polygynandrous) multi-female groups. Two factors may thus have caused gibbon females to disperse (van Schaik and van Hooff 1983): (a) strong female–female competition for food; (b) a reduced predation risk due to the extremely arboreal lifestyle of gibbons, their capability of high-speed brachiatory locomotion, and the absence of large primate-hunting eagles (such as crowned hawk eagle, *Stephanoaetus coronatus*; harpy eagle, *Harpia harpyja*: great Philippine eagle, *Pithecophaga jefferyi*) in their habitat.

When scanning the literature on the extent of predation risk actually faced by wild gibbons, only scarce evidence emerges: a full grown siamang was found in the stomach of a python (Schneider 1906) and gibbon hair in two out of 237 samples of leopard faeces (Rabinowitz 1989). When wild gibbons were caught in canopy cage traps during a malaria study in Malaysia, pythons invariably surrounded them, obviously trying to prey on the gibbons (van Gulik 1967).

So far, no act of predation has been observed during any gibbon study (Carpenter 1940, Ellefson 1974, Gittins 1979, Grether *et al.* 1992, Kappeler 1981, Leighton 1987, Mitani 1990, Palombit 1992, Raemaekers and Chivers 1980, Whitten, 1980). This includes the long-term study site at Khao Yai where the Mo Singto gibbon population (more than a dozen groups) has been monitored for more than 20 years (cf., Brockelman *et al.* 1998, unpublished data W.Y. Brockelman, U. Reichard, N. Uhde), despite the presence of at least nine species of potential predators at this site (see below). However at Mo Singto, gibbons of all age and sex classes have disappeared despite the virtual absence of animal poaching in this area. Some of these disappearances might well stem from unwitnessed cases of predation (unpublished data U. Reichard, N. Uhde).

The present chapter intends to investigate how gibbons cope with the risk of predation, particularly since gibbons forfeit the safety benefits provided by multi-female groups. First, we look at how predation risk and gibbon foraging heights are linked with each other. We then examine whether gibbons employ additional behavioral strategies to minimize predation risk beyond the degree of safety they already attain by their arboreal lifestyle and the absence of large eagles in their habitat. For this, our study can rely on a compilation of anecdotal evidence to describe antipredatory behavior in the context of potential or actual predator–prey interactions which is unprecedented for gibbons. Finally, we discuss the male role in predation avoidance in gibbons.

Potential predators of gibbons at Khao Yai

Adult white-handed gibbons weigh 4.4–7.6 kg; infants of about 2 kg start to travel independently (Roonwal and Mohnot 1977). At the Khao Yai study area (see below), several species may potentially prey on such relatively lightweight primates:

1. *Cats* use optical and auditory senses when hunting. They include clouded leopard (*Neofelis nebulosa*, 11–23 kg), marbled cat (*Pardofelis marmorata*, 2–5 kg), Asian golden cat (*Catopuma temmincki*, 8.5–15 kg), leopard cat (*Prionailurus bengalensis*, 2.5–5 kg) and tiger (*Panthera tigris corbetti*, 100–195 kg). The occurrence of leopards (*Panthera pardus*, 29–70 kg) in Khao Yai is disputed. The clouded leopard and marbled cat are specifically adapted to arboreality. Even the relatively heavy tigers are surprisingly agile in ascending tree trunks. Cats usually hunt by either stalking or 'sit-and-wait,' before launching an ambush attack. They are mostly active during crepuscular hours but

sometimes during the day as well (Gittleman 1989, Jackson and Nowell 1996, Kitchener 1992, Lekagul and McNeely 1977, unpublished data S. Austin). Daytime sightings of active clouded leopards, leopard cats, marbled cats and tigers in the study area support this (Davies 1990, Nettelbeck 1993, Raemaekers and Raemaekers 1990, unpublished data NLU, VS). Clouded leopards have been reported to feed on arboreal primates such as proboscis monkeys in Borneo (for review, see Jackson and Nowell 1996).

2. *Snakes.* Khao Yai is inhabited by reticulated pythons (*Python reticulatus*) who can become larger than any other snake. Prey detection relies predominantly on smell, apart from optical and infra-red sense. In the study area, pythons were seen to hunt muntjak deer (Nettelbeck 1995) and pig-tailed macaques (personal communication A. Intaruk). Pythons tend to search for prey actively during the night but shift to a 'sit-and-wait' strategy during the day. They hunt both on and above the ground (Bauchot 1994, van Schaik and Mitrasetia 1990, Whitten 1980).

3. *Raptors.* Asian birds of prey are not large enough to pose a serious threat to adult gibbons but infants and juveniles meet the size and weight features of several raptors (Brown and Amadon, 1968). At least four species of raptors at Khao Yai are strong enough to take young gibbons: changeable hawk eagle (*Spizaetus cirrhatus*, 56–75 cm), mountain hawk eagle (*Spizaetus nipalensis*, 66–75 cm), black eagle (*Ictinaetus malayensis*, 69–78 cm), and crested serpent eagle (*Spilornis cheela*, 51–71 cm). Hawk eagles hunt predominantly from inside the forest via 'sit-and-wait.' The two other species tend to fly over the forest and dive into the trees during surprise attacks (Brown and Amadon 1968, Lekagul and Round 1991).

Predation risk and foraging

For gibbons, the safest locations are high up in the trees. Body mass limits the diameter of branches, trunks and thus heights that cursorial predators such as cats and pythons can climb. Visibility increases with canopy height and so increases the probability of early predator detection (van Schaik *et al.* 1983). Birds of prey hunt from both inside and outside the forest at any given canopy height (Brown and Amadon 1968). Hence, raptor avoidance suggests avoidance of exposed parts rather than avoidance of great heights. As a consequence, gibbon safety from predators increases with canopy height.

Gibbons are specialized on fresh ripe fruit (Gittins and Raemaekers 1980), which is most abundant high up since light is an important factor in fruiting. Most gibbon fruit trees are fairly tall when they finally reach maturity and start to bear fruit (Whitten 1980). Ripe figs, the keystone resource of gibbons, are usually found at rather great heights. In contrast, the lower canopy heights yield few edible items for gibbons, except for insects and occasional leaves, shoots and fruits of climbers (Whitten 1980).

Do gibbons take predation risk into account while they forage? We would expect them to avoid the forest floor and to forage at heights that exceed average canopy height. Second, we expect a group with more fully grown animals and no infant to forage more likely at risky areas and thus at lower heights.

Antipredatory behavior

Gibbons do not reduce predation risk by living in large groups, which is typical for many other diurnal primate species. It is further unlikely that gibbons employ a cryptic antipredatory strategy during their activity period, as has been suggested for De Brazza guenons (Gautier-Hion and Gautier 1978), considering the gibbons' highly conspicuous singing, play and locomotion. Apart from safety in numbers or crypsis, animals might reduce predation risk by, for example, spatio-temporal avoidance, early detection, escape, mobbing and deterrence (Caro 1995, Harvey and Greenwood 1978). Do gibbons employ additional behavioral strategies to minimize predation risk?

Spatio-temporal avoidance

Since for gibbons, safety increases with height (see above), under the predation hypothesis we predict that they avoid the ground and select for great heights whenever possible. A group with more fully grown animals and fewer immatures will be more likely to use lower heights than others. One might argue that elevated positions might be chosen to facilitate early spotting of conspecifics. This would predict higher positions in the home range periphery than in its center, because at Khao Yai, the greatest social threat comes from neighboring groups rather than from roving strangers (NLU unpublished data). One might speculate that gibbons select for high locations while performing activities with reduced opportunity for vigilance. We would then expect a correlation between the vigilance level and the height for the different activities.

Early detection, escape and mobbing
This antipredatory strategy predicts adult gibbons to be alerted
easily by noises that might indicate the presence of a potential
predator such as alarm calls of other species or uncommon sounds.
Furthermore, adult gibbons are expected to react strongly towards
potential predators whereas to remain indifferent when encountering relatively harmless animals (interactions with food competitors
or birds defending their nest site represent an exception to this; see
also Whitington 1992).

Behavioral adjustment after potential predator encounter
Predators often stay for several days in an area before they move on.
Therefore, gibbons might modify their behavior for the next few
days once they obtain evidence of a potential predator's presence.
Safety might be increased by a higher group cohesion, a lower
feeding activity and by the use of greater heights. Close spatial proximity can improve safety through, for example, the effects of dilution and earlier detection (Hamilton 1971, Landeau and Terborgh
1986, van Schaik *et al.* 1983). Reduced feeding activity could give
more time for activities that allow greater vigilance such as sit-and-
hang.

The male role in predation avoidance

A close female–male association and special male services can play
a substantial role in antipredator strategies (van Schaik and van
Noordwijk 1989). Male service has been hypothesized to be crucial
in the evolution of monogamy, once gibbon females started to live
dispersed. A male could increase his fitness through predator monitoring, enabling the female to feed more safely (Dunbar 1988). This
would predict male gibbons to be more vigilant than females.

Methods

Data collection

The study was conducted in the tropical moist evergreen forest of
Khao Yai National Park, Thailand (2.168 km², 130 aerial km NE of
Bangkok; 14°N 101°E). The population of white-handed gibbons
(*Hylobates lar*) in the Mo Singto study area (800 m above sea level) is
believed to be at carrying capacity (four to five groups per km²) and
at least 12 groups have been monitored since 1978 (for further
information, see Brockelman *et al.* 1998).

Table 16.1. *Gibbon study groups: composition and observation sample Jan–May 1996*

Group (Scans)	Sex	Age class	Age in years[1]	Name	Individual scans[2]
A	F	Adult	>28	Andromeda	464
(722)	M	Adult	~22	Fearless[3]	496
	M	Subadult	8.5	Amadeus	461
	M	Juvenile	5.5	Aran	507
	F[4]	Infant	2.5	Akira	456
B	F	Adult	>30	Bridget	474
(689)	M	Adult	13	Chet[5]	502
	M	Adult	>30	Bard	463
	F	Subadult	8.5	Brenda	393
	F	Juvenile	4.5	Benedetta	471
C	F	Adult	>23	Cassandra	425
(647)	M	Adult	>26	Cassius	501
	M	Subadult	9.5	Christopher	301
	F[4]	Juvenile	5.5	Caleb	424
	F[4]	Infant	2.5	Cyrana	422

Notes:

[1] Date of reference: 15 March 1996.

[2] Reflect how often a given individual could be detected during the group scans.

[3] Transferred from group F into group A in 1983.

[4] Individual labelled as M by mistake in some previous papers (e.g., Reichard and Sommer 1997, Brockelman *et al.* 1998).

[5] Transferred from group C into group B during 1994.

The data on activity, height above the ground and group cohesion were derived from a study conducted by NLU from January to May 1996. Three well habituated groups of gibbons (groups A and C: three fully grown gibbons, one juvenile, one infant; group B: four fully grown gibbons, one juvenile, no infant) were directly observed for 538 h using *ad-libitum* and scan sampling at 10-min intervals, yielding 2058 group scans, which were equally distributed over the day (see Table 16.1). The data on adult gibbon vigilance levels were recorded by NLU during Dec 1999–Feb 2000 on four habituated groups (A, B, C and T, one adult female + one adult male). Qualitative data on gibbon reactions towards unusual events and other species were collected by NLU (1993–2000) and VS (1995–96) as well as provided by other researchers (mostly from 1992 to 1996).

Observations of all groups using 8×30 Zeiss binoculars began at dawn and ended between 15:00 h and 16:00 h when the gibbons settled in their night trees. Scans started with the individual located farthest to the left side if more than one gibbon was in view. Within one vertical line, the animal at the higher spot was recorded first. The following activity classes were recorded:

- *Doze.* Lying with open or closed eyes or sitting upright with eyes closed.
- *Play.* Autoplay and social play. Movements, wrestling and chasing while at least one gibbon shows play-face and/or laughs.
- *Groom.* Allo-grooming involving two or more individuals.
- *Locomotion.* Move (brachiate, climb, leap, walk) within a tree or from tree to tree. All breaks of movement were scored as sit-and-hang.
- *Feed.* Reach for, process, swallow and masticate food items.
- *Sit-and-hang.* Sit or hang on a substrate while all other categories can be excluded.
- *Sing.* Vocalize loudly, alone or with others.

Height above the ground was estimated in meters. Accuracy was trained with an optical height measurer.

Group cohesion was defined as the percentage of individuals in a scan who had a neighbor at <5 m. In order to minimize errors estimating distances at different canopy heights, a gibbon's armlength was taken as additional reference.

Vigilance. An animal was recorded as vigilant when it had its head up and, within 15 s after the scan beep, either stared intently into one direction for more than 5 s or looked around with a head movement of at least 45°. Inspection of vegetation or partners at a close range was not scored as vigilance.

Gibbons at Khao Yai utter various alarm vocalizations (similarly described by Ellefson, 1974): (a) *Brief alarm call* ('bark hoot;' high-pitched, short single note). (b) *Disturbance call* ('hoo-sigh;' medium to high pitched two-phase vocalization produced by drawn out exhalation and inhalation). (c) *Series alarm calling* ('hoot series;' which may culminate in full-blown solo songs, duets or group singing).

Data analysis

We reduced the data set to one averaged data point per individual and calculated 95% confidence intervals for the average height above the ground during different activities. Vigilance levels were calculated as the percentage of scans the animals were found vigilant. We used the Mann–Whitney-U test (Siegel and Castellan 1988) to investigate changes in the gibbon behavior after the encounter with a tiger.

A 10-min interpoint-sample interval maximizes the time between samples while still ensuring that scan sampling provides a representative sample of the time devoted to common activities. In this study, scans with an interval of 10 min were not independent of each other (chi-square test: $\chi^2 = 97.65$, df$=9$, $p<0.001$) whereas statistical independence was achieved when using a 20 min interval ($\chi^2 = 12.17$, df$=9$, NS). For this reason, we omitted every second scan when comparing vigilance and activity patterns of females and males with a chi-square test. The same applied when we used a chi-square test to examine group differences in the use of height above ground.

Results

Activity, group size and height above ground

The gibbons at Mo Singto spent almost 90% of their feeding time at heights between 10 and 40 m (Fig. 16.1). Average foraging height (24.7 m, SD 1.7 m; A 26.5 m, B 23.1 m, C 24.5 m) showed no deviation from average canopy height at Mo Singto (25.5 m, SD 12.5 m; smallest vertical range containing 50% of height measurements: 20–34 m, cf., Brockelman 1998). There was no difference between groups in the amount of time spent feeding at less than 20 m above the ground (A 29.4%, B 36.8%, C 35.1%; chi-square test: $\chi^2 = 4.97$, df$=$ 2, NS; group B: four fully grown gibbons, one juvenile, no infant; groups A and C: three fully grown gibbons, one juvenile, one infant).

As with other activities, during foraging the gibbons normally avoided the forest floor. However, exceptions were observed. Gibbons came to the ground for seconds to catch large insects, for example grasshoppers or stick insects. On a few occasions they would briefly feed on climber fruits, leaves or shoots at less than 2 m above the ground. Furthermore, the gibbons occasionally came to the forest floor during escalated intergroup encounters and during play of juveniles and subadults. Such behavior is rarely

Fig. 16.1. Height above ground during feeding. Groups A (black bars), B (white) and C (grey) in comparison.

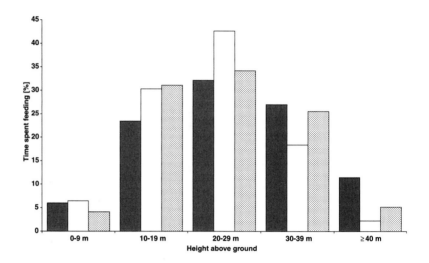

reported for these quintessential arboreal apes because previous studies did not achieve the necessary degree of habituation towards human observers. Increased safety associated with humans is an unlikely cause for ground movements of gibbons since the gibbons were sometimes already on the ground when observers arrived.

Height above ground varied for different activities (Fig. 16.2a–c). The greatest canopy heights were chosen when dozing (1.2% of daytime activity), playing (5.5%), grooming (11.8%) or singing (4.1%). Activities such as locomotion (20.1% of daytime activity), feeding (25.1%) and sit-and-hang (27.3%) were performed at lower heights.

The heights used for locomotion, feeding, sit-and-hang might reflect food distribution and forest structure. In contrast, the gibbons chose greater heights for doze, play, groom and sing than expected from average canopy height. For example, 50.1% of all social grooming occurred in 30–39 m. Moreover, canopy heights refer to treetops whereas grooming occurred mostly in the middle or lower tree sections (49.1% and 28.4%, respectively), probably to avoid sunshine during hot hours (most grooming occurred after 10:00 h). There was no difference in grooming heights between the core area of the home range (average 30.7 m, SD 2.8 m; A 31.3 m, B 27.7 m; C 33.1 m; $n = 67$ grooming sessions) and areas that overlap with neighbors (average 27.7, SD 2.4 m; A 26.6 m, B 30.4 m, C 26.1 m; $n = 126$ grooming sessions).

(a)

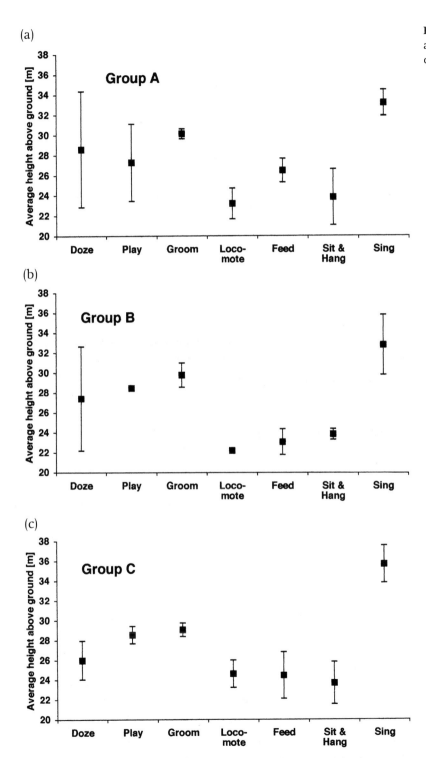

Fig. 16.2. Average height above ground during different activities.

(b)

(c)

Fig. 16.3. Overall height above ground. Groups A (black bars), B (white) and C (grey) in comparison.

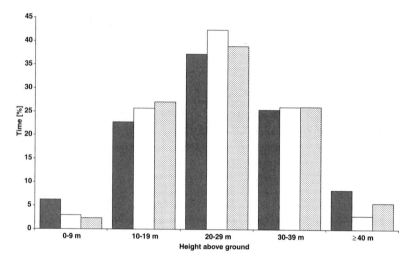

There was no relationship between the average height of an activity and the average vigilance level of this activity. Vigilance was high during sing (92.3%) and sit-and-hang (87.2%). During all other activities, vigilance was greatly reduced (locomotion: 18.2%; groom: 16.1%; feed: 11.2%; doze: 0.7%; play: 0.0%). Adult males spent more time with sit-and-hang than adult females (females 30%, males: 35%; chi-square test: $\chi^2 = 9.12$, df $= 1$, $p < 0.01$) and less time feeding (females 28%, males 21%; $\chi^2 = 19.13$, df $= 1$, $p < 0.001$). However, adult females were as vigilant as adult males (females: 49.5%, males 51.3%; $\chi^2 = 0.85$, df $= 1$, NS).

The overall height average was similar for groups A (25.7 m), B (25.0 m) and C (25.5 m); see Fig. 16.3. There was no difference in the amount of time the groups spent lower than 20 m above the forest floor (A 29.0%, B 28.7%, C 29.4%; $\chi^2 = 0.27$, df $= 2$, NS).

Interspecific reactions

Both adult and immature gibbons were easily alerted by alarm calls and mobbing noises of other species (Table 16.2). For example, gibbons feeding a few meters above the ground immediately jumped higher and stared intently to the forest floor when a barking deer gave sudden alarm calls in close proximity. On 21 April 1996, continuous alarm calls of a barking deer farther away in their home range elicited series alarm calling by group C that culminated in group singing. The group abruptly ended an encounter with group A and moved quickly towards the barking deer, while group A stayed at the encounter area and gave a few disturbance calls.

The gibbons were alerted and gave brief alarm calls following hornbill or brief squirrel alarm calls which are typically associated with a bird of prey that glides low over the tree tops. Squirrels, hornbills or small birds may utter continuous alarm when a bird of prey or a snake has been detected within a tree; this would frequently elicit mobbing by the gibbons.

Uncommon noise associated with the locomotion or vocalization of seemingly alerted large animals or other potential indicators of a predator's presence (e.g., rocks rolling down a slope, thick branches falling down) would also cause alarm in the gibbons. In contrast, barks of an Asiatic black bear evoked no gibbon reaction. It is unlikely that former poaching caused these reactions because the study groups are habituated to noise made by humans, particularly tourists.

The presence of relatively harmless animals (some of which were food competitors, e.g., binturongs, pig-tailed macaques, squirrels) was almost always ignored by adults. They only elicited alarm calls or mobbing by the less experienced immatures. In contrast, the Khao Yai gibbons reacted strongly when encountering potential predators. Pythons, large birds of prey and a tiger provoked alarm calls and intense mobbing by adults, including loud calling, branch shaking and slapping the potential predator. For the most part, adult males took a more active role in mobbing than adult females.

Large birds of prey which flew low over the trees invariably caused the gibbons to give brief alarm calls and immatures to drop from the tree crown or crouch towards the trunk. Groups without infants (group B, this study) or without any immature members (group T, unpublished data, NLU) also gave alarm calls once a raptor was spotted. The gibbons seemed to take the degree of imminent danger into account because they did not react to large birds of prey that flew high above the forest. They showed no reaction towards the Khao Yai National Park helicopter, which sometimes flew over at less than 40 m.

Reaction towards a tiger

An uncommon observation involves a gibbon–tiger encounter: '11 May 1996, 09:46 h. The adult male of group A utters disturbance calls, followed by solo singing. Series alarm calls and group singing

Table 16.2. Interspecific reactions of gibbons in the Khao Yai study area

Category	Species	No. of Events	Reaction[1] (A=Alarm, M=Mob, F=Flight, I=Ignore)		Observer[2]
			Adults	Immatures	
Encounter potential predator					
Tiger	*Panthera tigris*	1	A, M	A, M	NU
Python	*Python reticulatus*	6	A, M	A, M	WB, AN, UR, NU
Large raptor (sit in tree)	*Ictinaetus malayensis, Spilornis cheela, Spizaetus cirrhatus, Spizaetus nipalensis*	2	A, M	A, M	UR, NU
Large raptor (fly over low)	(as above)	15	A	A?, F	UR, VS, NU
Large raptor (fly over high)[3]	(as above)	>25	I	I	NU
Encounter other animal					
Elephant	*Elephas maximus*	1	I	I	UR
2 Malayan sunbears	*Helarctos malayensis*	1	I	I	AN
15–20 wild pigs	*Sus scrofa*	1	I	I	NU
Binturong	*Arctictis binturong*	7	A	A, M	VS, NU
Monkey	*Macaca nemestrina*	>25	I	A, M	VS, NU, CW
Squirrel	*Callisciurus finlaysoni, Ratufa bicolor*	>25	I	M	VS, NU
Other snake	*Trimesurus sp.,* other	3	I	A, M	VS, NU
Small raptor (defend nest)	*Accipiter badius*	>10	A, F	A, M, F	AN, NU
Large hornbill (fly over low)	*Buceros bicornis, Rhyticeros undulatus*	>50	I	I	NU, US
Small bird	e.g., Columbidae, *Anthracoceras albirostris*	7	I	M	VS, NU

Alarm calls other species

Alarm calls barking deer	*Muntiacus muntjak*	>25	A, M, F	A, M, F	RR, NU
Alarm calls squirrel (brief)	*Callosciurus finlaysoni*	>50	A	A?, F	RR, NU
(continuous)		>30	A, M	A, M, F	RR, VS, NU
Alarm calls hornbill	*Anthracoceras albirostris*	>30	A	A?	NU
Alarm calls jungle fowl	*Gallus gallus*	1	A	–	NU
Other event					
Barks of Asiatic black bear	*Selenarctos thibetanus*	1	I	I	NU
Unknown large animal(s) run, bark, scream		2	A, M, F	A, M, F	NU
Uncommon noise (e.g., fall of rock or thick branch)		>15	A, M, F	A, M, F	RR, VS, NU

Notes:

[1] A, alarm and/or disturbance calls (includes interruption of activity and increased vigilance); M, mob and/or harrass (may include approach the animal/source of noise, follow, loud calling, locomotional display, branch shaking, hitting, chasing); F, flight reaction and/or seek cover (may include jump to a higher position, drop from substrate, move towards adult gibbon or center of treetop; not included were flights caused by retaliation of an animal harassed by the gibbons).

[2] Unpublished data (1989–2000) from: WB, Warren Brockelman; AN, Anouchka Nettelbeck; UR, Ulrich Reichard; VS, Volker Sommer; US, Udomlux Suwanvecho; NU, Nicola Uhde. Previously published data: RR, Raemaekers and Raemaekers 1990; CW, Whitington 1992.

[3] More than 50–100 m above the forest.

follow after the adult female joins in. Quick arboreal travel brings the gibbons back into the direction they just came from (the sub-adult male in front, followed by the juvenile male and the others) and from where three loud growls are heard at 09:56 h. At 09:58 h, a tiger is spotted by the observer (NLU) about 50 m away at the location the gibbons are heading to.' Backward travel is highly uncommon (apart from a cul-de-sac situation, which did not apply). Thus, the gibbons moved obviously intentionally towards the tiger. The observer left the spot but recorded prolonged continuous singing of group A for about 105 min in roughly the same area.

Cohesion of group A was highest (48%) during the observation block following the tiger encounter (Fig. 16.4a), although the small sample size limits the statistical examination of this result (Mann–Whitney U test: $U=0$, $n_1=1$, $n_2=5$, $p=0.167$). Feeding activity was lowest after the encounter ($U=0$, $n_1=1$, $n_2=5$, $p=0.167$; Fig. 16.4b). The time spent in less than 10 m above the ground was not related to the tiger encounter ($U=2$, $n_1=1$, $n_2=5$, $p=0.500$; Fig. 16.4c).

Discussion

Predation risk and foraging ecology

Gibbons at Khao Yai normally avoided the forest floor. This is unlikely to be an observer artifact, since the Mo Singto gibbons are by far the best habituated population of wild gibbons, and some animals may approach observers as close as 1 m. The gibbons foraged in high forest layers compared to other primates as has been found at other sites (Gittins 1983, Whitten 1980). Thus, they are safer from predators than, for example, the sympatric pig-tailed macaques who mostly feed and travel terrestrially or less than 10 m above the ground.

However, gibbon foraging patterns seem to reflect merely food distribution rather than avoidance of predation risk. Despite the fact that gibbon foraging heights are great relative to other primates, average foraging heights at Khao Yai only matched average canopy height, a result similarly found at other sites (Chivers 1977, Gittins 1983, Gittins and Raemaekers 1980, Whitten 1980). Gibbon group B with more fully grown individuals and no infant foraged at the same canopy heights as the other groups did instead of exploring more risky areas as expected by the predation hypothesis. Still, further studies that investigate how group size, cohesion

(a)

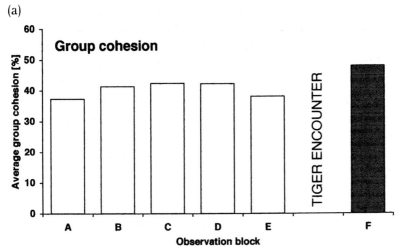

(b)

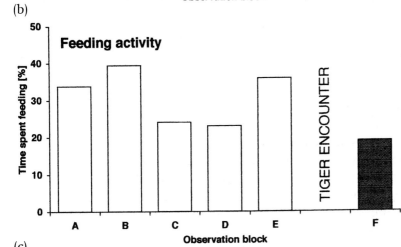

(c)

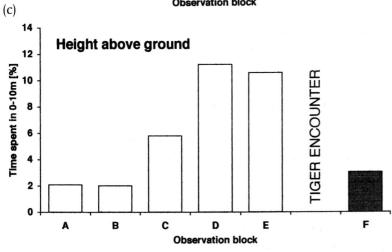

Fig. 16.4. Gibbon behavior before (white bars) and after (black bar) encounter with a tiger. (a): Group cohesion. (b): Feeding activity. (c): Time spent in less than 10 m above ground (blocks: A = 2, 4 and 5 Feb 96; B = 6, 7 and 8 Feb 96; C = 12, 13 and 17 Mar 96; D = 20, 23, 30 and 31 Mar 96; E = 15, 16, 18, 19, 21 and 22 Apr 96; F = 12, 13 and 14 May 96).

and foraging heights relate to feeding success are needed to clarify a possible influence of predation risk on gibbon foraging decisions.

Negligible predation risk for gibbons?

The foraging pattern of gibbons, the lack of large eagles in their habitat, their ability of high-speed locomotion and the probability of detecting predators from a large distance (since visibility in the forest increases with height; van Schaik *et al.* 1983) lead to a reduced predation risk in gibbons at relatively small group sizes (cf. van Schaik and van Hooff 1983). Some authors described this residual risk as 'virtually absent' or 'negligible' (Carpenter 1940, Ellefson 1974, Grether *et al.* 1992, Leighton 1987, Raemaekers and Chivers 1980). However, we disagree with this judgment for two reasons. (a) The loss of only a single infant compromises the reproductive output of a gibbon female more than that of a female cercopithecine because birth intervals in Hylobatidae are longer than previously thought and do approach those of other apes (>3 years; cf., Brockelman *et al.* 1998, Mitani 1990, Palombit 1992). Gibbons can live for decades (at least 44 years in captivity; Mootnick and Sheeran 1995; the adult female of group A was probably >35 years old) and have a long juvenile life-stage, raising the costs of predation far above those of shorter-lived species. (b) Our observations suggest that gibbons employ potentially costly strategies to minimize the remaining predation risk (cf., Ellefson 1974, Raemaekers and Raemaekers 1990). We elaborate on this second point below.

Antipredatory behavior

Spatio-temporal avoidance

As predicted, the gibbons usually avoided the forest floor. Rare exceptions were related to food (see above), escalated intergroup conflicts or play. A gibbon, when chased at a critical close range, invariably flees downwards and even onto the ground, presumably to avoid a cul-de-sac situation in the treetops. Play of juveniles and subadults on the ground might have an important function for young gibbons to become familiar with an environment which they might have to turn to later during serious fights (NLU unpublished data).

The lower height above ground (relative to other activities) during behaviors related to foraging (feed, locomotion, sit-and-hang) seemed to reflect forest structure (average canopy height 25.5 m), food distribution and energetic aspects; for example, traveling via the top of the trees would increase travel costs immensely. In contrast, for grooming, playing and dozing, higher and thus safer spots were selected. Social grooming and play are often associated with a digestive break by the entire group. Singing was also performed high up, probably to ensure broadcasting efficiency. Other studies have found a similar pattern in the use of heights (Chivers 1977, Gittins 1983, Gittins and Raemaekers 1980, Whitten 1980).

One might argue that these elevated grooming positions are chosen to facilitate early spotting of conspecifics. However, this would predict higher grooming sites in the home range periphery than in its center, which was not the case. One might speculate that gibbons choose high locations while performing activities with reduced opportunity for vigilance. Since high vigilance levels were found only during singing and sit-and-hang, this presumption could not be supported. Thus, the gibbons seem to select greater heights if they have a choice to do so, as during the 'leisure' activities groom and play. There was no difference between groups with or without infants concerning the heights above ground they selected.

Our study focused on daytime behavior. However, gibbon sleeping habits could likewise reflect temporal predation risk avoidance. For example, gibbons break up into subgroups in the afternoon and select particular sleeping trees (Reichard 1998). These trees are less likely to have thick branches interwoven with neighboring trees which makes it more difficult for cursorial predators to sneak up to sleeping gibbons (NLU unpublished data).

Early detection and escape
The Khao Yai gibbons were easily alerted by uncommon noises or signs of the presence of a potential predator, for example alarm calls of other species. They would then interrupt their activity, increase alertness, utter disturbance calls, and sometimes move towards the source of the noise (similarly noted by Ellefson 1974, Kappeler 1981, Raemaekers and Raemaekers 1990). Several flight reactions by immatures in response to large birds of prey were observed. Gibbon groups without infants or without any immatures also gave alarm calls once a raptor was spotted. It is possible that these alarm responses are too firmly rooted to show any variation according to current group compositions.

Mobbing and deterrence
Mobbing by gibbons at Khao Yai included loud vocalization, pursuit of the potential predator and even chasing it off as in the

case of raptors that sat in a tree. Adult Khao Yai gibbons clearly discriminated between potential predators and relatively harmless animals. The latter were only mobbed by immatures who may lack experience or use a harmless target to 'practice' mobbing. At Ketambe (Sumatra), prolonged and intense alarm calling by siamangs and long-tailed macaques was observed for more than 2 hours before dusk against a clouded leopard (van Schaik personal communication). On four occasions, lar gibbons in Malaysia vocalized loudly for a long time immediately after a tiger had roared (Ellefson 1974).

Mobbing and loud calls of adults may inform other group members, especially immatures, about the presence of a potential predator as well as its location and may indicate their willingness to defend the offspring. However, predator mobbing can also be observed in primate groups that consist largely of nonrelated individuals (such as Hanuman langur male bands; Rajpurohit *et al.* 1995). Therefore, mobbing does not necessarily reflect kin-relations but may signal to predators that they have been detected, that the mobbers can afford to reveal their own presence, and that it might be better to avoid the area of an unsuccessful hunt in the future (Caro 1995, Harvey and Greenwood 1978, 'handicap-principle' sensu Zahavi and Zahavi 1996).

A case study: behavior after a tiger encounter
In gibbon group A, cohesion was highest after the tiger encounter, as predicted. Feeding activity, as expected, was lowest during the days after the encounter. Contrary to our prediction, the time group A spent less than 10 m above ground was unaffected by the tiger. Van Schaik and Mitrasetia (1990) similarly found increased group cohesion but no changes in height above ground in long-tailed macaques after the encounter with a model python. Experiments with models of predators such as stuffed snakes, raptors and cats as well as play-backs are needed to gain more insight into gibbon patterns of reaction. A study by Treves (1999) indicates there might be a considerable variation between species in the response to predator encounters.

The male role
A male vigilance service that relieves the female of predator monitoring has been hypothesized to be crucial in the evolution of monogamy (Dunbar 1988). Our data from Khao Yai do not support this idea for gibbons. Although adult male gibbons at Khao Yai will

mob most actively, there was no difference in vigilance levels between males and females. Moreover, males are often far away from female and offspring – in fact, they travel mostly *behind* them. Finally, males are frequently distracted by intergroup encounters. Therefore, other forces have likely brought about permanent female–male associations in gibbons, such as infanticide or resource defense (Reichard and Sommer 1997, van Schaik and Dunbar 1990). Protection from predation should therefore be viewed as a secondary benefit.

Predation risk and group size

Our results suggest that gibbons employ antipredatory strategies to further minimize their relatively low predation risk. The increase in safety gained by these strategies must be less costly than the safety increase that could be obtained by forming multi-female groups. There are two alternatives: (a) Gibbons can 'cheaply' minimize the predation risk to a very low level even at small group sizes. Thus, the need to form multi-female groups does not arise. (b) Minimizing predation risk at a small group size is costly but still less so than the formation of larger groups because the benefits derived from larger groups do not outweigh the costs arising from female–female competition for food. Further studies are needed to clarify these cost–benefit trade-offs.

Acknowledgments

We thank the National Park Division of the Royal Forestry Department, the National Research Council of Thailand, the former and present superintendents of Khao Yai National Park (Mr Vallobh Sakont, Mr Phairat Tharnchai, Dr Chumpon Sukkasame) who kindly granted research permission to NLU and VS. NLU thanks the Thai–Danish Training Center and the Khao Yai National Park staff for support and hospitality, especially Mr Amnuay Intaruk, Mr Prayuth Lorsuwansiri, Mr Saroej Prapan, Ms Duanghatai Reangkesa, Mr Sanya Sorralum, Mr Somkiat Susunpoontong, and Ms Sompong Yousuk. We thank Warren Brockelman, Center for Conservation Biology, Mahidol University, Bangkok, for support and encouragement. Sean Austin, Warren Brockelman, Anouchka Nettelbeck, Ulrich Reichard, Carel van Schaik and Udomlux Suwanvecho kindly allowed use of unpublished data. The Abteilung Verhaltensforschung and Ökologie, Deutsches Primatenzentrum, provided facilities for data analysis which was kindly assisted by

Domingo Mendoza. We thank Carola Borries, Andreas Koenig, and Carel van Schaik for comments on an earlier draft. We thank Claire Bracebridge and Mark Read who helped to correct the English. NLU thanks the Heinrich Böll Stiftung for financial support.

REFERENCES

Bauchot, R. (ed.) (1994). *Les Serpents.* Chamalières: Atp, Bordas.

Brockelman, W.Y. (1998). Study of tropical forest canopy height and cover using a point-intercept method. In: F. Dallmeier and J.A. Comiskey, eds., *Forest Biodiversity Research Monitoring and Modeling: Conceptual Backround and Old World Case Studies, Man and Biosphere*, Vol. 20, UNESCO, New York: Paris and Panthenon Publishing, pp. 521–31.

Brockelman, W.Y., Reichard, U., Treesucon, U., and Raemaekers, J.J. (1998). Dispersal, pair formation and social structure in gibbons (*Hylobates lar*). *Behavioural Ecology and Sociobiology* **42**: 329–39.

Brown, L.H., and Amadon, D. (1968). *Eagles, Hawks, and Falcons of the World.* Feltham: The Hamlyn Publishing Group.

Caro, T.M. (1995). Pursuit-deterrence revisited. *Trends in Ecology and Evolution* **10**: 500–3.

Carpenter, C.R. (1940). A field study in Siam of the behavior and social relations of the gibbon (*Hylobates lar*). *Comparative Psychology Monographs* **84**: 1–212.

Chivers, D.J. (1977). The feeding behaviour of Siamang (*Symphalangus syndactylus*). In: T.H. Clutton-Brock, ed., *Primate Ecology: Studies of Feeding and Ranging Behaviour in Lemurs, Monkeys and Apes.* London: Academic Press, pp. 355–82.

Davies, R.G. (1990). Sighting of a clouded leopard (*Neofelis nebulosa*) in a troop of pig-tailed macaques (*Macaca nemestrina*) in Khao Yai National Park, Thailand. *Natural History Bulletin of the Siam Society* **28**: 95–6.

Dunbar, R.I.M. (1988). *Primate Social Systems.* London: Croom Helm.

Ellefson, J.O. (1974). A natural history of white-handed gibbons in the Malayan Peninsular. In: D.M. Rumbaugh, ed., *Gibbon and Siamang*, Vol. 3: *Natural History, Social Behavior, Reproduction, Vocalizations, Prehension.* Basel: Karger, pp. 1–136.

Gautier-Hion, A., and Gautier, J.P. (1978). Le singe de Brazza: une stratégie originale. *Zeitschrift für Tierpsychologie* **46**: 84–104.

Gittins, S.P. (1979). The behaviour and ecology of the agile gibbon (*Hylobates agilis*). Ph.D. thesis, Cambridge: University of Cambridge.

Gittins, S.P. (1983). Use of the forest canopy by the agile gibbon. *Folia Primatologica* **40**: 134–44.

Gittins, S.P., and Raemaekers, J.J. (1980). Siamang, lar, and agile gibbons. In: D.J. Chivers, ed., *Malayan Forest Primates: Ten Years' Study in Tropical Rain Forest*, New York: Plenum Press, pp. 63–105.

Gittleman, J.L. (ed.) (1989). *Carnivore Behaviour, Ecology, and Evolution.* Ithaca, New York: Cornell University Press.

Grether, G.F., Palombit, R.A., and Rodman, P.S. (1992). Gibbon foraging decisions and the marginal value model. *International Journal of Primatology* **13**: 1–17.

Hamilton, W.D. (1971). Geometry of the selfish herd. *Journal of Theoretical Biology* **31**: 295–311.

Harvey, P.H., and Greenwood, P.J. (1978). Anti-predatory defense strategies: some evolutionary problems. In: J.R. Krebs and N.B. Davies, eds., *Behavioural Ecology*. Oxford: Blackwell Scientific Publications, pp. 129–51.

Jackson, P., and Nowell, K. (1996). *Wild Cats. Status Survey and Conservation Action Plan*. Gland: IUCN.

Kappeler, M. (1981). The Javan silvery gibbon (*Hylobates lar moloch*): ecology and behaviour, Part II. Ph.D. thesis, Basel: University of Basel.

Kitchener, A. (1992). *The Natural History of the Wild Cats*. London: Croom Helm.

Kleiman, D.G. (1977) Monogamy in mammals. *Quarterly Review of Biology* **52**: 39–69.

Landeau, L., and Terborgh, J. (1986). Oddity and the 'confusion' effect in predation. *Animal Behaviour* **34**: 1372–80.

Leighton, D.R. (1987). Gibbons: territoriality and monogamy. In: B.B. Smuts, D.L. Cheney, R.M. Seyfarth, R.W. Wrangham and T.T. Struhsaker, eds., *Primate Societies*, Chicago: Chicago University Press, pp. 135–45

Lekagul, B., and McNeely, J.A. (1977). *Mammals of Thailand*. Bangkok: Association for the Conservation of Wildlife.

Lekagul, B., and Round, P.D. (1991). *A Guide to the Birds of Thailand*. Bangkok: Saha Karn Bhaet.

Mitani, J.C. (1990). Demography of agile gibbons (*Hylobates agilis*). *International Journal of Primatology* **11**: 411–24.

Mootnick, A.R., and Sheeran, K.L. (1995). Lar gibbon (*Hylobates lar*). In: W. Beacham, ed., *Beacham's International Threatened, Endangered and Extinct Species*. Washington, DC: Beacham.

Nettelbeck, A.R. (1993). Zur Öko-Ethologie freilebender Weißhandgibbons (*Hylobates lar*) in Thailand. Diplom Thesis, Hamburg: University of Hamburg.

Nettelbeck, A.R. (1995). Predation on barking deer by reticulated python and dholes in Khao Yai National Park. *Natural History Bulletin of the Siam Society* **43**: 369–73.

Palombit, R.A. (1992). Pair bonds and monogamy in wild Siamang (*Hylobates syndactylus*) and white-handed gibbon (*Hylobates lar*) in Northern Sumatra. Ph.D. thesis, Davis: University of California.

Palombit, R.A. (1994a). Extra-pair copulations in a monogamous ape. *Animal Behaviour* **47**: 721–3.

Palombit, R.A. (1994b). Dynamic pair bonds in hylobatids: Implications regarding monogamous social systems. *Behaviour* **128**: 65–101.

Preuschoft, H., Chivers, D.J., Brockelman, W.Y., and Creel, N (ed.) (1984). *The Lesser Apes: Evolutionary and Behavioural Biology*. Edinburgh: Edinburgh University Press.

Rabinowitz, A. (1989). The density and behavior of large cats in a dry tropical forest mosaic in Huai Kha Khaeng Wildlife Sanctuary, Thailand. *Natural History Bulletin of the Siam Society* **37**: 235–51.

Raemaekers, J.J., and Chivers, D.J. (1980). Socio-ecology of Malayan forest primates. In: D.J. Chivers, ed., *Malayan Forest Primates: Ten Years' Study in Tropical Rain Forest*, New York: Plenum Press, pp. 279–316.

Raemaekers, J.J., and Raemaekers, P. (1990). *The Singing Ape. A Journey to the Jungles of Thailand*. Bangkok: Amarin Printing Group.

Rajpurohit, L.S., Sommer, V., and Mohnot, S.M. (1995). Wanderers between harems and bachelor bands: Male Hanuman langurs (*Presbytis entellus*) at Jodhpur in Rajasthan. *Behaviour* **132**: 255–99.

Reichard, U. (1998). Sleeping sites, sleeping places and pre-sleep behavior of gibbons (*Hylobates lar*). *American Journal of Primatology* **46**: 35–62.

Reichard, U. (2001). Social grouping patterns in wild gibbons. *XVIIIth Congress of the International Primatological Society, 7–12th January 2001,* Abstract, Adelaide.

Reichard, U., and Sommer, V. (1997). Group encounters in wild gibbons (*Hylobates lar*): agonism, affiliation, and the concept of infanticide. *Behaviour* **134**: 1135–74.

Roonwal, M.L., and Mohnot, S.M. (1977). *Primates of South Asia: Ecology, Sociology and Behaviour*. Cambridge: Harvard University Press.

Schneider, G. (1906). Ergebnisse zoologischer Forschungsreisen in Sumatra. *Zoologische Jahrbücher. Abteilung Systematik, Geographie und Biologie der Tiere* **23**: 1–172.

Siegel, S., and Castellan, N.J. (1988). *Nonparametric statistics for the behavioral science*. New York: McGraw-Hill.

Sommer, V., and Reichard, U. (2000). Rethinking monogamy: the gibbon case. In: P.M. Kappeler, ed., *Primate Males: Causes and Consequences of Variation in Group Composition*, Cambridge: Cambridge University Press, pp. 159–68.

Treves, A. (1999). Has predation shaped the social systems of arboreal primates?. *International Journal of Primatology* **20**: 35–67.

van Gulik, R.H. (1967). *The Gibbon in China*. Leiden: Brill.

van Schaik, C.P. (1983). Why are diurnal primates living in groups? *Behaviour* **87**: 120–44.

van Schaik, C.P. (1989). The ecology of social relationships amongst female primates. In: V. Standen and R.A. Foley, eds., *Comparative Socioecology. The Behavioural Ecology of Humans and Other Mammals*, Oxford: Blackwell Scientific Publications, pp. 195–218.

van Schaik, C.P., and Dunbar, R.I.M. (1990). The evolution of monogamy in large primates: a new hypothesis and some crucial tests. *Behaviour* **115**: 30–62.

van Schaik, C.P., and Mitrasetia, T. (1990). Changes in the behaviour of wild long-tailed macaques (*Macaca fascicularis*) after encounters with a model python. *Folia Primatologica* **55**: 104–8.

van Schaik, C.P., and van Hooff, J.A.R.A.M. (1983). On the ultimate causes of primate social systems. *Behaviour* **85**: 91–117.

van Schaik, C.P., and van Noordwijk, M.A. (1989). The special role of male cebus monkeys in predation avoidance and its effect on group composition. *Behaviour, Ecology and Sociobiology* **24**: 265–76.

van Schaik, C.P., van Noordwijk, M.A., Warsono, B., and Sutriono, E. (1983). Party size and early detection of predators in Sumatran forest primates. *Primates* **24**: 211–21.

Whitington, C. (1992). Interactions between lar gibbons and pig-tailed macaques at fruit sources. *American Journal of Primatology* **26**: 61–4.

Whitten, A.J. (1980). The Kloss gibbon in Siberut rain forest. Ph.D. thesis, Cambridge: University of Cambridge.

Wrangham, R.W. (1980). An ecological model of female-bonded primate groups. *Behaviour* **75**: 262–300.

Zahavi, A., and Zahavi, A. (1996). *Handicap Principle: A Missing Piece of Puzzle.* New York: Oxford University Press.

Index